NEW TECHNOLOGY AND REGIONAL DEVELOPMENT

New Technology and Regional Development

Edited by
BERT VAN DER KNAPP
and
EGBERT WEVER

CROOM HELM
London • Sydney • Wolfeboro, New Hampshire

© 1987 G.A. van der Knapp and E. Wever
Croom Helm Ltd, Provident House, Burrell Row,
Beckenham, Kent, BR3 1AT

Croom Helm Australia Pty Ltd, Suite 4, 6th Floor,
64-76 Kippax Street, Surry Hills, NSW 2010, Australia

British Library Cataloguing in Publication Data

New technology and regional development.
    1. Regional economics 2. Technological
innovations — Economic aspects
I. Knapp, G.A. van der II. Wever, Egbert
330.9 HT38 HT391

ISBN 0-7099-3106-9

Croom Helm, 27 South Main Street,
Wolfeboro, New Hampshire 03894-2069, USA

Library of Congress Cataloging-in-Publication Data

New technology and regional development.

    Includes index.
    1. Technological innovations. 2. Regional planning.
3. Research, industrial. I. Knaap, G.A. van der.
II. Wever, E.
HC79.T4N48 1987   338.9′26   86-23983
ISBN 0-7099-3106-9

We wish to express our gratitude to Amat Geerdink Wordprocessing Service
for typing the manuscript.
Printed and bound in Great Britain
by Billing & Sons Limited, Worcester.

CONTENTS

Contents

Since the second half of the seventies, a period of economic recession, a growing interest for technological change can be noticed in all 'old' industrial countries. The reason for this phenomenon is the conviction that the economic future of these countries will depend to a large degree on their ability to create new products and processes and to make these commercially viable. Hence the stimulation and subsidising of all kinds of Research and Development (R&D) activities by the governments of these countries.

The strong emphasis in economic policy on technological change or innovation was accompanied and supported by scientific research. Although policy measures strongly favoured technological research, many investigations dealt with the relation between technological change and economic development. In this type of research some 'old' ideas were revaluated, as was the case with the Kondratiev long waves, Schumpeter's ideas about entrepreneurship and the concept of product life cycle.

Within this line of research geographers together with some economists took a specific position, as they are especially interested in questions of a spatial nature. In what kind of regions are conditions most favourable to realise technological change? Will the process of technological change lead to growing differences between regions and what policy measures can be taken to prevent such further divergence? What will be the impact of spatial differences in the process of technological change in general?

These kinds of questions raised the interest of not only geographers, but also of planning authorities. This certainly has to do with the fact that technological change manifests itself in very specific locations. Just as technological change may be associated with Silicon Valley, with Route 128, with Sophia Antipolis or Cambridge University, this, at the same time, introduces a spatial element in the discussion.

Against this background it was certainly not unexpected that the IGU Commisson on Industrial Change, presided by Godfrey Linge, decided to organise a congress on 'Technology and Industrial Change.' This congress was held in August 1985 at Nijmegen in the Netherlands. In this book some of the papers presented there are published. We are very happy that we were allowed to include the very interesting contributions of the representatives of Philips in Eindhoven and of the European Commission in Brussels.

We should like to place on record our appreciation of the help afforded by the Department of Geography and Planning of the Catholic University of Nijmegen and by the Economic Geographic Institute of the Erasmus University of Rotterdam. Sincere thanks must also go to the Royal Dutch Shell, for sponsoring part of the costs for preparing the camera-ready text.

Rotterdam   Nijmegen
Bert van der Knaap   Egbert Wever

CHAPTER ONE

TECHNOLOGY AND INDUSTRIAL CHANGE: AN OVERVIEW

Dr. G.A. van der Knaap, Professor Economic Geographical
Institute, Erasmus University, Rotterdam and National Physical
Planning Agency, The Hague, The Netherlands[1]
Dr. G.J.R. Linge, Professorial Fellow Department of Human
Geography, The Australian National University, Canberra,
Australia
Dr. E. Wever, Professor Department of Geography and Planning,
Catholic University, Nijmegen, The Netherlands.

Introduction

Since 1970 the world economic environment has experienced
considerable changes. This has become manifest through frequent
alternations in currency exchange rates, interest rates and
commodity prices. The latter have led industrial organisations to
make decisions increasingly frequent with respect to changes in
investments, stocks, wages and labour requirements (Hamilton
1984). These trends converged and intensified to make global
restructuring a necessity towards the end of the 1970s. These
changes are especially noticeable in the traditional capital
goods industries such as transport, building, factory equipment,
shipbuilding and steel industry.

The direction of this global shift, and its orientations
and speed, is not very clear. In the early 1970s it was thought
that the shift was from the North Atlantic Basin towards the
South Atlantic and as a result the centre of economic growth
could shift towards Latin America. This view has been supported
by the very rapid growth of the economies in these countries,
especially in Brazil. The dualistic North-South model is now
considered a very static and oversimplified view of the real
changes which take place (Forbes 1984). The countries in South
and Southeast Asia experienced a rapid development of their
economies, which led to a change from import-substitution
policies to export-led policies. High inflation rates and
associated high interest rates in combination with huge
petrodollar surpluses created after 1973 in the OPEC countries
have also shifted the focus towards South and Southeast Asia. In
this region are several of the Newly Industrialising Countries
(NICs), as well as Australia and Japan on the west side of the
Pacific Basin. Together with the western part of the United
States and Canada these countries have accounted for half the
increase in world production during the last five years. A major
growth factor in the NICs of this region has been the low wage
cost which increased their international competitiveness, as few

1

wage-related expenses (sick pay, holiday pay, compensation for work-related injuries) which were becoming the norm in industrialised countries were (or still are) payable there. The relative advantage seems to have weakened since the beginning of the 1980s. The countries in Western Europe in particular have improved their position by achieving a considerable drop in relative wage costs. This was realised through a successful wage restraint policy and associated low inflation rates. In contrast, increasing labour costs in South and Southeast Asia, especially in Singapore, may slow down, at least temporarily, the shift from the North Atlantic towards the Pacific Basin.

Within Europe the major shifts in sectoral employment which started already in the 1950s have continued at an accelerated rate. Both in agriculture and in industry there has been a growth in output concomitant with a considerable loss in employment. The service sector remains the last major growth sector in employment opportunities. Although the short term dramatic effects in terms of severe unemployment are felt heavily in a number of occupational and age groups, these changes will probably create the necessary conditions for economic growth without employment growth. The first indications of such a development are already noticeable for manufacturing, construction and wholesale activities. This is a situation which most countries in Northwest Europe are likely to experience after the year 2000 in light of the demographic changes that occurred after the second half of the 1960s. Low birth rates and zero population growth are likely and an absolute decline after 2025 is possible.

Yet there is an entrenched view which pervades the literature that Western Europe and North America are, and will remain, the economic core of the world, and what is happening elsewhere is something of an aberration. Most of the arguments can be considered a product or a derivative of the neoclassical paradigm that North American and West European economies are - in their normal state - closed systems in equilibrium (Weinstein and Gross 1984). The essence of the discussion is about the ways in which to restore this equilibrium. There is, as Weinstein and Gross note, a spectrum of arguments within this paradigm. At one extreme is the view that industrial change is a crisis, that the contraction of basic industries, and the related dislocation, is evidence of the inability of the economic system to effect a transfer of productive resources from declining to emerging industries equitably and efficiently, and that therefore there is strong case for government intervention including central planning and nationalisation of industry and other activities. At the other extreme is the view that high levels of unemployment and the decay of basic industries is the result of government mismanagement and interference, i.e. largely the failure of Keynesian demand-management policies. Inherent in this view is that the modification of government policies is the key to restoring the economy to a 'normal' state.

An alternative paradigm gaining creditability in the

literature is the antithesis of the first since it views economies as open systems in dynamic disequilibrium. This paradigm has a well established provenance in the Kondratieff and Schumpeter long-wave and business cycle theories published in the 1920s and 1930s. In the last two decades there has been an increasingly large number of papers on aspects of the life cycle of firms and products and more recently on innovation and investment wave cycles. Also within this paradigm there is room for arguments about intervention and no intervention. Those opposing intervention argue that it distorts price signals which allocate productive resources between competing opportunities, i.e. that market forces should prevail. Those accepting the need for intervention make allowances for the fact that not all the actors involved can adapt to change at the same rate because of financial, legal and other institutional circumstances over which they have no control. This implies, that they would accept forms of intervention that are aimed at reducing the frictions involved so that change is facilitated, but not intervention aimed at preventing change and maintaining the status quo. Inherent in this dynamic disequilibrium view is the notion that existing patterns of economic activity and trade and the way they are organised in space will be constantly breaking up and rearranging in different shapes and combinations.

Out of these observations arise a number of questions related to technology and industrial change and their spatial impacts. These questions are centred around the theoretical concepts being used to examine industrial change, the mechanisms of change, and the related type of (causal) effects on spatial arrangements, organisational structure, and the changing relations between industries and their various environments.

The analysis of industrial change is pursued from a variety of theoretical concepts which do not form one coherent structure. Most of these analyses are limited because they are based upon a rather traditional definition of industry, as manufacturing only and not other activities. This ignores the fact that industrial activities refer to specific types of production processes, which can be better described by a new classification recognising that industrial processes can be more accurately described by type of function and type of activity within a business organisation or enterprise (Törnqvist 1973) than by the division of the economy into three sectors, i.e. primary, secondary and tertiary activities. This classification, which may have been useful about 50 years ago (Fisher 1935, Clark 1940), is orientated towards employment structure. As the number of jobs in the primary and secondary industries is declining, employment figures are becoming less relevant in revealing industrial change. Other factors related to the work processes, the role of technology and the importance of the environment of the firm should be closely examined to understand the nature of the changes taking place. Especially important in this context is a reconsideration of the extent to which traditional location theories can still be used to explain and understand the locational behaviour of firms and enterprises.

The Role of Technology

Technological change is not something that suddenly appears. It is an integral part of human activities since the occupation of the earth's surface. The interesting part of technological change is that it also creates both institutional and social change. Thus technological change can be viewed as an instrument for the reorganisation of one's environment. However, technology is not an unequivocal determinant for direction but only constitutes a condition. Although most authors present only their own perspective, the possibility of steering technological development is based upon a greater freedom of choice created by technological innovations.

This freedom of choice may not be equally accessible at the same rate for everyone in society but it does create new opportunities. This argument is fully explored by Naisbitt (1984) in his recent book 'Megatrends'. In this study he presents society as a 'multiple option society', which he works out into ten major social developments for the future. This view is also shared by Toffler (1980, 1983) who strongly emphasises the increasing pluralism in society. In his view there is a greater freedom of choice of residential forms and working conditions made possible by telecommunications. In addition, there is a greater variety of goods as a result of more flexible production processes. A crucial element in this discussion is the use of advanced micro-electronics in the production of goods and services. Related to this is the introduction of modern organisation systems. The combined effect of these two changes may lead to a different use of space. Space requirements per unit of production or service may decrease sharply. This does not imply that in the event of strong economic growth there will be no great demand for new space. One of the problems associated with the application of micro-electronics and telematics is that very often they require a special environment which is usually not available in the existing structures. Thus new specially developed sites will be selected in favour of existing ones which were formerly used for additional production processes.

The impact of the application of telematics can be very dramatic in terms of the additional distinction between 'place of work' and 'place of residence'. In Toffler's (1980) view, telematics make it possible to work at home and still maintain sufficient contact with the parent firm. Inspired by the type of environment necessary to create the sufficient conditions, he speaks of the 'electronic cottage'. Such a development could have a number of dramatic effects, not only on the labour market participation of men and women but also on the social structure of society. Moreover if 10 to 20 per cent of the work was performed 'at home' this would have significant consequences for the structure of the economy and the spatial organisations of cities. Even if the idea of working at home in this particular way is not acceptable, say for psychological reasons, there is still sufficient material left to argue in favour of the emergence of a home-centred society and the development of neighbourhood economies (Olsen 1982). At present this may seem to be a picture of a very distant future, which could or could not happen after the turn of the century. A crucial element in this discussion is the type of technology adapted, the speed at which these technological innovations are diffused, and the sector of the economy in which they are adopted first.

Technological change is not referring to a single type of technology but to technology in its compound sense. This view is related to a definition of technology as 'the body of knowledge that is applicable in the production of goods'. In this sense technological change is a broader concept than technical change, which refers to a way of doing things. According to Adikibi (1985), there are three types of technology relevant for the production of goods:

a. organisation technology;
b. operative technology;
c. proprietary technology.

Organisation technology includes the skills and the techniques for organising plant lay-out and managing activities in the enterprise, i.e. the traditional staff functions. Operative technology is concerned with the skills and techniques for the physical operation and the maintenance of machinery, plant and equipment. Proprietary technology is generally firm specific and it is mainly derived from R&D activities and closely associated with a specific project, product or process. This latter type constitutes the core know-how of the production process. It is the central type of technology upon which the firm's survival depends. Against the background of the spread of technology this distinction can be relevant to ownership, control and regional effects especially in the case of large multinational corporations. Both organisation technology and operative technology can be implanted but a dependency relationship will remain with the parent firm and its home region. In the case of the spread of proprietary technology the situation differs as this is the critical technology. A different relationship is created between the corporation from which it originates and the recipient organisation or plant, which may eventually lead to a considerable degree of independence. In that case it is more appropriate to describe this diffusion process as the transfer rather than the transplant of technology.

Another distinction which seems to be relevant is the effect a technological innovation has on the organisation and the production of goods. Technological innovations which affect the efficiency of the production process e.g. by creating conditions for gaining further economies of scale are relatively neutral to their environment, unlike those that change the capital intensity of production or affect substitution between production factors. A good example of such an innovation is the introduction and perfection of flexible automation systems and flexible production systems in which by means of computer-controlled machines (industrial robots) considerable savings in space can be realised through a reduction of stocks (Marandon 1980. See also Grotz in chapter six). An important characteristic of this change is the shift from 'economies of scale' towards 'economies of scope'. This observation does not imply that economies of scale are no longer relevant, but it emphasises that the organisation of production is becoming more important.

The distinction made above into three types of technology can be compared with the classical distinction of innovation into product innovations and process innovations (Schumpeter 1939). This leads to a two-way classification in which type of technology and type of innovation are matched. Kok et al. (1984) go one step further and suggest a classification into four levels of innovations, i.e. basic, primary, secondary and tertiary. Basic innovation is something completely new which has a general economic impact. Primary innovation according to these authors is an innovative development related to a basic innovation and can in this respect be considered as new to the world. Secondary innovation is not entirely new, but it is new to the region, whereas tertiary innovation is an adaptation or an improvement of something already in existence and operative in the region. This classification is used by Kok and Pellenberg in chapter ten.

As basic innovations of this kind are rare, the main interest will be on the spatial and organisational implications of the other three types of innovations. Basic innovations are often proprietary technology and are the result of the R&D activities of large corporations. The other types can also be found in the small and medium-sized independent firms. Considering the diffusion of information (about innovations) between these firms, three models of explanation are used by various authors:
a. the incubator theory of the urban economists (Pred 1966);
b. the product life cycle theory as formulated by Vernon (1966);
c. the concept of hierarchical diffusion within an urban system (Brown 1981). This diffusion of information may create the appropriate conditions for the transfer of technology.
The incubator theory or seedbed growth concept provides a link between urban and industrial economics. The theory is based upon the role agglomeration economies and localisation economies (Isard 1956) play to provide an attractive environment for the growth of new innovative firms located in the centre of the city. The theory is based upon historical evidence: yet few contemporary studies of the growth of the innovative firm do support this view.

Another attractive idea is the product life cycle theory which is static in nature despite its process orientated appearance. In this theory there are four stages each associated with a particular stage in the growth of a firm. The attractiveness of this theory for geographers and regional economists is the spatial interpretation of these stages (Norton and Rees 1979). The first stage is characterised by experimentation and product development, which is skill intensive and labour extensive. There is little demand for space and it can be accommodated in a single plant. In this stage there is a link with the incubator theory because the centre of the large urban agglomerations can provide the right location for these kinds of activities. The second stage is the strong growth of a successful product. There is an increasing need for larger production units occupying greater space, more capital and good management. In this stage agglomeration economies and skilled labour are still relevant production factors. A good location is a spacious site at the fringe of a large urban agglomeration. In the third stage there is a stabilisation and a standardisation of the production process. The availability of capital and cheap labour is more important than skilled labour and agglomeration economies. Hence this stage can be used for the relocation of this type of activities from a core region to a peripheral location or an international relocation. This may be considered to be the filtering down process as described by Myrdal (1957). The fourth or final stage is the reduction of production sometimes followed by a closure.

There are problems associated with this conceptual scheme (see also Taylor in chapter five). The first is that the model is in essence static although it creates the impression that it is a dynamic model. In its regional application it has not been tested on the basis of its underlying micro behavioural assumptions. On a macro geographical level none the less, it may be useful as a descriptive tool. A problem arises at the micro level because it appears to be concerned with the time path of only one firm. On the aggregate, however, there is growth and decline of specific firms in specific categories at specific locations and this has not been linked with longitudinal observations for these firms. What can be concluded is that after some time there has been a transfer of technology from the core to the intermediate regions and the periphery. This view on the diffusion of innovations has been described in more spatial detail by Hägerstrand (1967) who considered the process entirely as an urban phenomenon. The transfer of technology follows the urban hierarchy which is related to functional distances constrained to some extent by physical distance based upon actual spatial location. In this way the urban centres in the intermediate zone and the periphery have, in descending order, the possibility of acting as hosts for a new firm.

A less rigid view on the urban structure is an urban system perspective with more complementarity and less hierarchical dominance. This allows for more locational flexibility. In a study carried out for the EEC Commission for Regional Policy, Keeble et al. (1981) have demonstrated on an EEC-wide scale that the major growth in industrial activity occurs in the intermediate zone. This observation raises many questions about the relationship between the innovative firm and the urban environment, about the relevant activity-space of the entrepreneur regarding his action, information or decision spaces, and about the type of industrial activity examined. This last comment refers back to the previous discussion about the appropriate classification of industry. Van der Knaap and Louter (1986) have recently demonstrated that innovative and promising activities in the commercial services have stronger ties to the main urban areas than innovative and growth activities in industry. This agrees with Keeble et al.'s observations at an EEC level as the growth in the intermediate zone is based upon the modern sectors of industry, whereas growth in the periphery is still occuring in the more traditional ones. However, it has to be realised that modern industrial sectors are closely associated with jobless growth, while the commercial services can at present be associated with growth in employment as is also the case with the expansion of traditional industries.

Although some observations have been made about the increase in freedom of locational choice both for the individual and for the firm, this statement certainly has to be clarified as to what people and to which firms it is applicable, the extent to which it is applicable and the nature of the new constraints that influence spatial decisions.

Importance of the Environment

With the relative decrease in the significance of traditional location factors, the role that the production environment plays in the organisation of industrial activities has gained in importance. The production environment in this context can be broadly interpreted as referring to the spatial environment, the organisational environment of the firm and the decision environment of the entrepreneur. This environment will vary with the size of the firm.

There is a considerable body of literature suggesting that large firms perform the lion's share of R&D (Malecki 1983). The implications of this observation for the small firm and its potential for growth is by no means clear. On the basis of traditional locational theory, this would be the kiss of death for the small firm, as the potential to innovate is the only way to survive. This may be the case for large firms but is not necessarily true for small ones. There are several arguments for a divergent view about the growth potential of large compared with small firms. Existing theories had until recently little to contribute towards the understanding of the complex geographical structure of the multi-plant firm (Hamilton and Linge 1979). The growth of the small firm into a large multi-plant firm within a regional context has been described by Watts (1980). The organisational structure of the firm can be understood in terms of its response to an increase in size, which allows corporate functions, such as R&D and marketing, to be largely separated from production. Whatever the structure of the firms, the location of the jobs controlled by individual plants are quite diverse and cannot be simply explained in terms of corporate structure (Pred 1977).

Another element related to the organisational structure of the firm is the strategy pursued dependent upon the level of technology applied within the firm. In this way strategy may not only become a means to distinguish between environment, but also between firms and between firms of different size. In the past few years there have been several interesting developments. The first is the growth in the number of small firms not functionally related to the multi-plant firm. The small firm has a considerable growth potential outside the core area and is very prominent in the intermediate and peripheral areas. According to Conti (1985), who examined evidence of the Italian periphery, this spatial redistribution of productive activities can neither be interpreted as an entirely endogenous or exogenous phenomenon nor as a consequence of a hierarchical diffusion process. It is based upon a valorisation of local resources and refers to a different organisation of production which cuts across traditional spatial patterns. The new development pattern is led by small local entrepreneurs, by family enterprises and by public administration itself who have attuned themselves to this form of industrial production. This example demonstrates that the interaction between social and cultural change can be a motive force towards industrial change, without the presence or existence of a new technology. It may, however, be argued that within the region itself it can be considered as a tertiary innovation (Kok et al. 1984) founded on an existing and known technology. Nevertheless in this case, the origin of the growth in the number of small plants is based upon local forces in response to local demand and not upon the existence of high level of R&D or the presence of highly innovative firms.

A development which is noteworthy too is the changing organisational structure of large firms both externally and internally. After a period of dynamic growth facilitated by expanding markets, take-overs and mergers this growth came to a halt in the second half of the 1970s and a process of restructuring began. The need for restructuring was caused by factors, such as market saturation and bad management, as the rate of growth was faster than internal adjustments could be made. The recognition of this situation has led to a spatial reorganisation associated with closures of some plants (Stafford and Watts 1985), while other sections of the corporate organisation regained greater independence through increased self-management. Partly simultaneous with this reorganisation of the external environment there is also the reorganisation of the internal structure, i.e. the work process itself. In light of the broad definition of technological change, this restructuring of the work process can be considered an organisational technology. An interesting example of such an organisational change is the shift from the Taylor system and the Ford system, which are based upon achieving direct economies of scale towards a group production technology as applied in the Volvo car industry in Sweden (Alvstam and Ellegard 1982 and 1985). This group technology involves a flexibility of management and allows for easy application of modern technology in situations where diversification leading to producing short runs and market fragmentation are becoming dominant characteristics.

In this type of production process the social structure of the labour process is increasingly important, as the role and the position of the individual employee is becoming a dominant factor in the changesin the production structure, which moves from economies of scale towards economies of scope. One of the consequences of such a development is a drastic change in the range and composition of employment, which will necessitate a different basis for the distribution of prosperity. The traditional link between technological adjustment, increase in value added, economies of scale and the wage negotiations seem to be becoming less evident. This type of change in the work process will also affect the relation between work and the traditional pattern of time consumption, and can provide the basis for what is often referred to as a 'leisure-time society' (Linge 1984). Thus it is not only the range and the composition of employment which are important when considering the possible impact of technological change, but also the quality of the job is a significant issue. There are already several indications that point towards a dichotomy of jobs (Schuman 1984). This evidence suggests that there has been a reduction of the number of jobs at the intermediate level, resulting in a relatively small group of highly qualified jobs and a large group of low qualified ones. Until now the distribution of income or welfare has been strongly linked to the performance of work. The question how to achieve an equitable distribution of income which is not based on paid employment will be one of the social challenges of the future. This requires a careful analysis and monitoring of the changing relations between income, work and employment: the latter can no longer be considered as synonymous with the other.

Some emphasis here has been put on the changing relations within the production process which, on a lower spatial scale can be associated with a fragmentation of the workplace. This may lead to various types of segmentation, e.g. a segmentation of production into large corporations and small firms, and a segmentation of the labour market into highly-qualified and low-qualified jobs. This raises many questions about power relations and the spatial distribution of particular segments of production and the labour force as a consequence of such relationships. According to Dostàl (1984), technological change and its inter-regional unequal employment implications is different between and within large and small corporate organisations because of an existing power asymmetry between large and small firms. Power, in this context, refers to the capacity and the ability of an organisation to regulate access to resources of other organisations and in this way force the latter into types of decision making which otherwise would not have occurred. To study these differences within a segmented economy, Taylor (1983) has proposed a classification of large and small firms into four types following the concept of the product life cycle: these are leaders, intermediates, laggards and supporters (this classification is used by Morphet in chapter three). Leaders are engaged in high-risk operations of the development of new products or markets requiring skilled labour, know-how and managerial expertise. This is part of the long-term survival strategy of the large corporation. Intermediate enterprises are producing reliable and continuing profits from established markets: they need managerial skills and sufficient capital. The laggard companies are often involved in standardised, sometimes obsolete manufacturing products near the end of their life cycle. Consequently the laggard will experience a fall in demand, a changing cost structure and a deskilling of its labour force. They seem to be characteristic of pheripheral regions.

Leading small firms very often have only a limited independence. As they are the result of personal initiatives and innovations and have a restricted financial basis they are likely to be taken over by large corporate organisations. Intermediate small firms are often older ones, and Taylor (1983) proposes that they be divided into two categories: the 'loyal opposition' and the 'satellites'. The latter are integrated in the production chains of a large scale corporation through subcontracting. The former are laggard small firms of the artisan and craftsman type which remain small as there is no ambition to expand. Such a classification can be extended (see Taylor and Thrift 1983) to include inter regional differences in labour market situations. These authors argue that inter-regional variations in innovations are a function of different mixes of firms as described above and their position in power and dependency relationships. Two general types of regional structure of firm mixes can be identified:

a. the structure of lagging regions, usually characterised by small corporate groups, branch plants of large corporations and non-innovative small firms;
b. the expansion and growth-orientated structure of core regions, characterised by effective innovative corporate organisations and a strongly innovative small firm sector.

Taylor and Thrift argue that this results in a cumulative process of regional decline in the periphery and a growth in the core region. However, as these two regional structures are linked through inter-regional systems, there should be a balancing mechanism operating that would slow down the decline in the one and the growth in the other region.

The evidence is by no means as clear-cut as may appear from the above described mechanism of regional change. Villeneuve and

Rose (1985), for instance, point out that there are at least three competing hypotheses from which the effect of technological change on the labour force can be evaluated. The first assumes an overall rise in skill levels (Aydalot 1976, Molle 1983). In complete contrast is the second proposed by Massey and Meegan (1982) and supported by Hyman and Price (1983), who foresee a deskilling of the labour force. A third, intermediate, position is taken by Lipietz (1980) who argues that a polarisation in occupational structure is most likely to occur. This polarisation tendency takes place in the occupational structure of certain branches and will lead to an increasing functional and spatial division of labour. Lipietz has called this process 'branch circuit' as the spatial division of labour no longer involves an entire branch but relates specifically to the different types of labour present within a branch. This may lead to a further spatial separation of the activities because of 'Fordism'. One may now add that with the introduction of 'neo-Fordism', being defined as the introduction of computerisation in labour processes (Villeneuve and Rose 1985), the simultaneous concentration of control and fragmentation of manual operations will be enhanced.

The main point about the three hypotheses is not that they cannot occur simultaneously, as they are closely related, but to determine which is spatially the most dominant. There is evidence from various studies that both reskilling and deskilling occurs. Spatial polarisation may be the outcome of the joint process of technological and occupational change, but this spatial result can only be properly explained from a careful analysis of the changing sets of causal relationships at the micro level. This involves not only studies from the demand side of the labour market, but also the supply side which should get more attention than it has received until now (Linge 1985).

Labour Market and Industrial Change

One of the main themes in the discussion about future patterns of industrial activities is the fundamental change from an electro-mechanical based technology to a micro-electronic based technology and it is the wide scope of this change which is at the root of many problems and uncertainties being faced today. One uncertainty is the extent of this conversion and the rate at which it will take place. As a particular case in point, a few years ago it was thought that teleconferencing would reduce the need for people to travel. In practice it has been found that face-to-face contacts and the ability to socialise are important considerations, while the development of faster trains has made travelling less time consuming.

Another important uncertainty is the impact of technological change on employment. In fact such a change has always created winners and losers. New and superior products displace old and inferior ones which lead to a loss of assets by some producers and a loss of jobs by some workers. The basic fear is that technological process will not result in more output for the same input, rather in the same output being produced with fewer inputs and, specifically, as has already been indicated above, with fewer inputs of labour. It is important to recognise however, that structural adjustments of this kind are not the product of technological change alone.

Attitudes to technological change and its impact on the demand for labour tend to fall into three categories (Browne 1985), although these are by no means clearly separate. In part this interrelatedness arises because of the distinction often made between a product innovation and a process innovation. It is the latter which some see as the greater threat to employment. Unfortunately, this concept - which has been widely promoted in the literature - is less useful and valid than it may appear at first sight, as the discussion in section 2 illustrates. Furthermore, a process innovation in one industry may be a product innovation in another: a new robot design is a product innovation for its maker but becomes a process innovation for its users.

The first attitudinal category embraces the view that has existed since the early nineteenth century when the Luddites in Britain destroyed machines which they thought would displace them. While it is recognised that innovation creates jobs it is feared that, overall, more jobs will be destroyed than created. However, this argument depends on three assumptions:
a. innovations are mainly of the process type;
b. process-innovations will reduce labour inputs, rather than capital or material inputs, per unit of output;
c. total output does not increase sufficiently to offset decline in labour inputs.

Against these assumptions can be put the view that product innovations can lead to a substantially higher value of output with the same or even more inputs. Much of the fear about fewer labour inputs being required stems from the fact that micro-electronics and data transmission and processing are at the centre of it all and that these are more process-orientated technologies than those of the past. As against this view, however, it can be argued that a pure process innovation accompanied by a reduction in labour inputs will lead to a decline in price per unit of output, and hence to an increase in demand which will at least partially offset the decrease in labour inputs. This may be considerable if, for example, the lower unit cost enables domestic producers to increase exports because they have become more competitive. In this case the benefits of introducing technological change accrue to the country concerned but at the cost of another country whose domestic producers are adversely affected. This, indeed, is why some who are concerned about the deleterious impacts of technological change on employment in fact support such a change. They fear that the employment impacts of not remaining internationally competitive would be even worse.

The second attitudinal category covers those who argue that technological change has little net effect on the overall demand for labour because losses in some sector will be offset by gains in others which are expanding in response to the growth of the economy as a whole. It is argued that, of necessity, growth can be boosted by government expansionary monetary and fiscal policies. There are at least three problems about this. First, there is a limit to the ways and the extent to which governments in capitalist and mixed economies can stimulate growth without creating the possibility of inflation and increased interest rates. In addition some would argue that government intervention is undesirable because it can distort market signals and lead to misallocation of resources. Second, even if job losses were balanced by job opportunities, the skills required in the expanding activities are unlikely to be the same as those that had been used in the declining activities. Thus people have to acquire new skills, which is not always easy especially for the older workers, and it takes time. Third, the locations of the expanding activities may be different from those of the declining ones. In France and Britain, for instance, it is not only the Paris and London region which are attracting research and development kind of activities. There is also a considerable growth in R&D along the Mediterranean coast from Toulouse to Marseilles and Grenoble. In England the growth of new activities is very noticeable along the M1 corridor away from London in the direction of Reading and Bristol as well as towards the whole of Southeast England. Even within urban areas the high technology activities tend to be located away from existing industrial zones in areas where modern buildings can be erected in attractive surroundings. This geographical problem raises a series of issues. Whereas organisations have the ability to close a plant almost literally overnight, most individual households cannot respond as quickly, especially if this means moving elsewhere. One set of constraints relates to the lack of information about employment, housing, schools and other amenities, as well as the opportunities and benefits available in other parts of the country or even elsewhere in the world. These considerations are greatly influenced of course by the particular stage in the life cycle of a household and the extent to which members are bound by educational, financial and social obligations or aspirations. A second set of constraints in people's mobility concerns the differential opportunities available to people having so-called 'good' or 'bad' jobs, which are discussed in the literature on segmented or dual labour markets. Organisations are more likely to offer managerial and technical staff transfer packages that include bridging loans, removal and relocation expenses. Process workers may at least be offered a job at another location but no financial assistance. To take advantage of re-employment opportunities a process worker may be caught in the blind of selling a dwelling in a weakening property market in a declining area and trying to buy one in a strengthening market elsewhere. A third set of constraints involves the ability of both working spouses (and possibly other members of the household) to find satisfying jobs at a mutually acceptable alternative location. However, it is not just the reduced ability of a household to move that is important but also the reduced necessity to do so - at least in many of the developed countries - because of the increasingly comprehensive social security systems (Anell 1981).

The third attitudinal category includes those who regard technological progress as being essential to prosperity and who emphasise that the development of new products stimulates demands. The wide spread use of motor vehicles, for instance, has created the need for a network of service stations and repair facilities, a market for auto accessories and the necessity for medical, legal, financial and other services. However, this is a much more complex question than it may at first seem, because it involves consideration of the various ways in which final demand by individuals and households is changing. These include the more frequent replacement of goods, the more rapid sequential employment of goods and the simultaneous use of goods (Linge 1984).

The direction of employment trends as a result of technological change is difficult to predict. The activities that have created the greatest concern are in manufacturing and services industries closely associated with it. The new technologies being introduced, especially robotics, are so productive that output per worker must rise dramatically over time with fewer people being required to produce a given output. So far, however, the predicted increases in productivity have not occurred, not even in countries like the United States or Japan. In fact, on the contrary, there has been a decline in the rate of productivity growth in most industrial economies. The reasons are not clear but it may be a sign that the potential of technology is growing more rapidly than the ability of employees and employers - and, indeed, the organisation of society as a whole - to adapt to it. What evidence there is about the impact of technological change on manufacturing activities suggests that job losses take place when productivity growth in a particular country is slow relative to other countries.

Implication for the Analysis of Industrial Change

The nature of technological change and its impact on industrial change is not clearly understood at this moment and raises all sorts of questions. These relate to the theoretical concepts being used, the causal effects of technological change and the impact it has on organisations and spatial structures, the reaction from the environment, organisations and people and the changing relations and the development of new strategies.

In the introductory paragraph of this chapter a brief sketch was made of the changing economic map of the world. Against this background a number of questions have been raised about appropriate definitions and conflicting empirical evidence in light of the theoretical concepts being used. This question is discussed in more detail by Thomas in chapter two, where he attempts to relate the various theoretical concepts. These concepts are the result of evidence from a variety of spatial environments which affect the decision environment and decision behaviour.

The role of technology as a vehicle for institutional and social change (see section 2) is viewed (in chapter three) from the demand-side by Morphet. This approach is in contrast with most studies which focus on the supply of technology and its related innovations. The transfer of technology once introduced is an important item to be studied at various spatial levels. The transfer of technology through time can be studied with the use of the product life cycle model, though in chapter five Taylor discusses a number of conceptual ambiguities associated with its use. The spatial transfer of technology can be studied using spatial diffusion models of the Hägerstrand type. It appears from empirical studies that this model is of limited use in its classical formulation. Several factors contribute to this view, such as the influence of regional policy on the creation of science parks in a large number of industrialised countries as the United States, England and West Germany, and their subsequent spatial impact. Another factor is the rather footloose character of high technology industries. The way in which government and industry react to these changes is discussed for the case of the Federal Republic of Germany by Schamp (chapter eight) who points out that functional distance is becoming increasingly important. This result is supported by two case studies for The Netherlands by Kok and Pellenbarg in chapter ten and Wever in chapter eleven. The latter also emphasises the effect of changes in spatial scale. According to his study most of The Netherlands can be considered just as a single urban field with a considerable freedom to locate nearly anywhere within this functional urban region. Changes in the use of space at the micro level within plants are discussed by Grotz in chapter six. From this study it is demonstrated that technological change when micro electronics is involved does involve reorganisation of production at the factory level.

The reaction of organisations on the observed changes for different types of spatial environments is becoming extremely relevant for understanding the direction of change in spatial organisation of society. In this way insights into the strategy of an organisation can be used as a means to distinguish between environments. The importance of these types of insights are presented by Harrington (chapter four) on the basis of a study of firms in the semi-conductor industry. At the other end of the spectrum is a view from a large multinational firm, i.e. Philips, and its locational behaviour in reaction to technological change and innovation summarised by one of its corporate directors,

Muntendam (chapter nine). Changes in spatial scales and changing spatial environments are also of relevance for regional policy makers, who can use these spatial dynamics in an attempt to influence future locations of activities to redress regional imbalances. In his discussion of the EEC regional policy Mathijsen places some emphasis (chapter seven) on the impact of the influence of physical distance and the role of telecommunication in the transfer of knowledge.

An important message emerges from all the studies with respect to traditional neo-classical location theory. A number of factors associated with the locational costs such as energy, transport, raw material and labour costs have changed their relative position, and some have lost their importance in favour of secondary location factors. These relate to the role of the entrepreneur and his behaviour and decision environment. Another important factor, also discussed in section 4., is the quality of the labour market with respect to the available skill levels and the ability to use them. In addition to this is the growing importance of the role of the household, not only in offering a specific supply of labour but also in the way the structure of the household limits its spatial mobility and thus effects the industrial system. Finally it is important to note the concern of a large number of studies for a possible polarisation of society caused by the mechanism of technological change and the reaction of political and business organisations to this change.

Note

1. G.A. van der Knaap wants to thank Drs. A. van Delft for the stimulating discussions during the preparation of chapter one of the annual report 1984 of the National Physical Planning Agency of The Netherlands, Technology and Physical Planning. Elements of that discussion have also found their way in this chapter.

References

Adikibi, O.T. (1985) Technology transfer to less-developed countries by multinational corporations: does it actually occur?, Paper presented at a meeting of the IGU Commission on Industrial Change, August, Nijmegen, The Netherlands

Alvstam, C.G. (1982) International trade in a changing environment - a demand for new theory, Paper presented at a meeting of the IGU Commission on Industrial Systems, Linköping, Sweden

Alvstam, C.G. and K. Ellegard (1985) People-production, international division of labour, Pilot case and methodology, Paper presented at a meeting of the IGU Commission on Industrial Change, Nijmegen, The Netherlands

Anell, L. (1981) Recession, the Western Economies and the Changing World Order, Pinter, London

Aydalot, P. (1976) Dynamique spatiale et développement inégal, Economica, Paris

Brown, L.A. (1981) Innovation diffusion: a new perspective, Methuen, New York

Browne, L.E., (1985) 'New technologies and employment: conflicting views and technological progress' Economic Impact, 1, 8-14

Clark, C. (1940) The Conditions of Economic Progress, MacMillan, London

Conti, S. (1985) Innovation,=modernization and locational changes in the Italian industrial system, Paper presented at a meeting of the IGU Commission on Industrial Change, Nijmegen, The Netherlands

Dostàl, P.F. (1984) 'Regional policy and Corporate Organizational Forms: some questions of interregional social justice' In: M. de Smidt and E. Wever (Eds), A profile of Dutch Economic Geography, Van Gorcum, Assen, 12-39

Fisher, A.G.B. (1935) The clash of progress and security, MacMillan, London

Forbes, D.K. (1984), The Geography of Underdevelopment: A Critical Survey Croom Helm, Beckenham

Hägerstrand, T. (1967) Innovation diffusion as a spatial process, University of Chicago Press, Chicago

Hamilton, F.E.I. (1984) 'Industrial restructuring: an international problem' Geoforum, 15, 349-364

Hamilton, F.E.I. and G.J.R. Linge (1979) 'Industrial systems' In: F.E.I. Hamilton and G.J.R. Linge (Eds) Spatial Analysis, Industry and the Industrial Environment: Progress in Research and Applications, Vol. 1, Wiley, Chichester, 1-23

Hyman, R. and R. Price (1983) The new working class? Whitecollar workers and their organizations, MacMillan, London

Isard, W. (1956) Location and Space-Economy, MIT Press, Cambridge, Mass.

Keeble, D.E., P.L. Owens and Chr. Thompson (1981) The influence of Peripheral and Central Locations on the relative development of regions, Dept. of Geography, Cambridge, England

Knaap, G.A. van der, and P. Louter (1986) De middelgrote steden, Economisch Geografisch Instituut, Rotterdam

Kok, J.A.A.M., G.J.D. Offerman and P. Pellenbarg (1984) 'The regional distribution of innovative firms in The Netherlands' In: M. de Smidt and E. Wever (Eds) A Profile of Dutch Economic Geography, Van Gorcum, Assen, 129-150

Linge, G.J.R. (1984) 'Industrialisation and the household' International Social Science Journal, 36, 319-39

Linge, G.J.R. (1985) Technology and Industrial Change, Paper
    presented at a meeting of the IGU Commission on Industrial
    Change, August, Nijmegen, The Netherlands
Lipietz, A. (1980) 'Interregional polarization and the
    tertiarisation of Society' Papers and Proceedings of the
    Regional Science Association, 44, 3-17
Malecki, E. (1983) 'Technology and Regional Development: a
    survey' The International Regional Science Review, 8, 89-125
Marandon, J.C. (1980) 'Die industrielle Flächenplanung im
    technologisch-organisatorischen Prozess der
    Industrieentwicklung' Mannheimer Geographische Arbeiten, 7,
    Mannheim
Massey, D. and R. Meegan (1982) The anatomy of job loss: the how,
    when and why of employment decline, Methuen, London
Molle, W. (1983) 'Technological change and regional development
    in Europe' Papers and Proceedings of the Regional Science
    Association, 52, 23-38
Myrdal, G (1957) Economic Theory and Under-developed Regions,
    Duckworths, London
Naisbitt, J. (1984) Megatrends, ten new directions transforming
    our lives, Warnes Books, New York
Norton, R.D. and J. Rees (1979) 'The product cycle and the
    spatial decentralization of American manufacturing' Regional
    Studies, 13, 141-151
Olsen, M.H. (1982) Remote office work, implications for
    individuals and organizations, Center for research on
    information systems, Working Paper Series no 25, New York
    University
Pred, A.R. (1966) The spatial dynamics of U.S. urban-industrial
    growth, 1800-1914, MIT Press, Cambridge, Mass.
Pred. A.R. (1977) City-systems in advanced economies, Hutchinson
    Press, London
Schuman, G. (1984) 'The macro- and micro-economic social impact
    of advanced computer technology' Futures, June, 260-286
Schumpeter, J.A. (1939), Business Cycles, 2 vols., Mc Graw-Hill,
    New York
Stafford, H. and M.D. Watts (1985) The role of technological
    change in plant closures, Paper presented at a meeting of
    the IGU Commission on Industrial Change, Nijmegen, The
    Netherlands
Taylor, M.J. (1983) 'Technological change and the segmented
    economy' In: A.Gillespie (Ed), Technological Change and
    Regional Development, Pion, London Papers in Regional
    Science, 12, London 104-117
Taylor, M.J. and N. Thrift (1983) 'Business Organization,
    Segmentation and Location' Regional Studies, 17, 445-465
Toffler, A. (1980) The Third Wave, W. Collins, New York
Toffler, A. (1983) Previews and Premises, W. Morrow and Comp.,
    New York

Törnqvist, G. (1973) 'Contact requirements and travel facilities: contact models of Sweden and regional development alternatives in the future' In: A.R. Pred and G. Törnqvist (Eds) Systems of Cities and information flows, two essays, Lund Studies in Geography, (B) no 38, Gleerup, Lund

Vernon, R. (1966) 'International investment and international trade in the product cycle' Quarterly Journal of Economics, 80, 190-207

Villeneuve, P. and D. Rose (1985) Technological Change and the spatial division of labour by gender in the Montreal metropolitan area, Paper presented at a meeting of the IGU Commission on Industial Change, August, Nijmegen, The Netherlands

Watts, H.D. (1980) The large industrial enterprise, Croom Helm, London

Weinstein, B.L. and H.T. Gross (1984) 'Grassroots industrial policy' Challenge, July/August, 52-55.

CHAPTER TWO

THE INNOVATION FACTOR IN THE PROCESS OF MICROECONOMIC INDUSTRIAL
CHANGE: CONCEPTUAL EXPLORATIONS

Dr. M.D. Thomas, Professor Department of Geography, University of
Washington, Seattle, USA

Introduction

Our primary sources of understanding concerning the processes of
economic change are the dominant neoclassical economic growth and
location theories. These, however, appear to be inadequate for
their explanatory roles. Persuasive arguments in support of this
point of view have been made by a number of notable scholars such
as Scott (1983a,b), Nelson and Winter (1977, 1982), Freeman
(1982) and Pavitt (1979). It seems that we need to develop a new
way of looking at the roles factors such as innovation and
technical change play in the processes which bring about scalar,
qualitative, structural and spatial changes in industries and
firms. These scholars, and many others, are already contributing
to the development of a useful set of new perspectives on the
explanatory mechanisms for these patterns of change.

One of the ways of improving our theories is by
strengthening their microfoundations (Rosenberg 1982, Weintraub
1979). We therefore should look, at the microeconomic level, for
a greater understanding of the processes responsible for the
contemporary dynamic patterns of industrial change (Thomas 1985).

As compared to other firms, innovative firms are those that
spend the most on research and development (R&D) relative to
their sales, and employ the highest proportion of technology-
oriented workers. These firms are found primarily in the so-
called 'high technology industries' and are characterised by
their relatively strong emphasis on activities related to the
process of technological innovation. In the development of
explanatory frameworks, I will use a number of key concepts from
the literature on the entrepreneur, innovation, the behavourial
theory of the firm and economic growth. Attention is primarily
focussed on the economic dimensions of the functional behaviour
of the 'entrepreneur' or key decision-maker(s).

Entrepreneurship is conceived as the behaviour of an
entrepreneur or entrepreneurs which results in the commercial
production of a product and/or process innovation. In innovative
high technology firms the entrepreneur may be a single person who

is, or a specific group of persons who are responsible for making the crucial investment and long range strategy decisions. This key individual or individuals are assumed to possess both 'innate unlearned' Schumpeterian entrepreneurial capabilities and managerial capabilities which manifest high levels of 'learned skills'. These capabilities vary from person to person and they also vary in the same person during the individual's life time.

The crucial investment and long range strategy decisions impact various dimensions of an innovative firm. I propose to focus attention on the impacts of these decisions on the firm's capacity, product and spatial dimensions. The investment and long range strategy decisions are the result of varying combinations of routine and non-routine behaviours carried out by or for the firm's key decision maker(s). When these crucial investment decisions bring about changes in the capacity, product and spatial dimensions of firms then industrial change takes place.

Each of the three dimensions of the firm has a number of attributes and only a few of these are discussed. With respect to capacity change within a firm attention is usually focussed on scalar attributes and their positive or negative implications for growth. Qualitative changes in capacity are also of interest. For example, changing capital/labour ratios imply changes in employment levels; deskilling and reskilling of employees as well as changes in economic efficiency. Both scalar and qualitative changes in the firm's product or products have clear connections to the process of industrial change and merit attention.

The spatial dimensions have a number of attributes, some of which are internal to the single establishment or to multiple constituent establishments of the firm. Other attributes, of greater interest at this time, are associated with the external rather than the internal environment of the firm. Examples are the dynamic scalar and qualitative attributes of various spatial contexts which are related to different kinds of decision making behaviours carried out by the firm's key decision makers; and the changing spatial tendencies of firms during the life of their innovations.

Contemporary Research

Very little past and present theorising about the processes of entrepreneurship and industrial change has been carried out in a spatial context. The inadequacies of a spatial theory concerning these processes at the microeconomic level have been well described by Nelson and Winter (1977, 1982). A recent comprehensive assessment of theories of regional development by Gore (1984) reveals the weak state of this body of spatial development theory which, in essence, attempts to provide a coherent explanation for the processes of regional industrial change. The greater poverty of useful theory, as one might expect, is even more challenging when we seek to explain the role of entrepreneurship in industrial change at both the level of the firm and with respect to firms in a spatial context (Thomas 1985). The inadequacies of the major neoclassical location and production theories have also received trenchant criticism in recent thought-provoking papers by Scott (1983a, b) and Storper (1985).

The last decade has witnessed, however, a strong increase in conceptual and empirical investigations of the relationships between technological change and regional or spatial industrial change. Singly, and sometimes together, geographers and economists are making a growing number of important interdisciplinary contributions to our understanding of these processes of change. For example, Steed (1982) has studied in Canada the industrial change implications of small and medium-sized innovative firms at the national and selected sub-national geographic levels. The growth patterns of a number of American and British high technology industries and small firms have been studied, in association with research and development (R&D) and investment cycles, at the regional level by Oakey (1983, 1984, 1985). Malecki (1979, 1980a,b,c, 1981, 1982) has explored the impact on regional industrial change of location distribution patterns of R&D establishments in the United States. The spatial diffusion of a number of process innovations involving selected high technology industries was surveyed by Thwaites, Oakey and Nash (1981), and by Thwaites, Edwards and Gibbs (1982) in Britain, and by Rees, Briggs and Hicks in the United States (1985).

In 1983 Brugman added a spatial perspective to the study of technology innovation in technology intensive industry in the United States. Her empirical evidence revealed a dramatic shift in the concentration of new electronic product generation from the North East quadrant of the U.S. to the western and south-western rim states. Inter-regional location change and the roles of corporate and business strategies in the United States semiconductor industry were investigated by Harrington (1983, 1985a,b) while Saxenian (1981, 1985) documented a number of intra-regional spatial consequences of industrial restructuring in Santa Clara County (Silicon Valley), California and Route 128 region in Massachusetts. Scott (1983b) investigated the organisation and the logic of location for the printed circuits industry in Los Angeles.

Glasmeier, Hall and Markusen (1983), Armington, Harris and Odle (1983) and Glasmeier (1985) have provided important evidence of spatial tendencies in United States high technology industries. Markusen (1984a,b,c) has provided important recent information and insights concerning the impacts of defence spending on the spatial and other characteristics of the high technology industries in the United States in general and in California in particular.

These investigations of various facets of the relationships between the processes of technological and industrial change at various geographic scales exemplify the growing importance of the spatial dimensions of these processes. These and similar studies also underscore a need for the articulation of a conceptual explanatory framework which will facilitate the identification, assessment and understanding of the economic, technical, political, social as well as the spatial dimensions of these processes.

Spatial-Temporal Context for Change in Innovative Firms

If we wish to understand the process of long term change in innovative firms, located in a specific macro or micro geographic region, then it is necessary to understand the dynamic nature and significance of the spatial and temporal context for effecting change in these firms. The designated place or region, like each individual firm, has a number of dynamic dimensions. For example, it has economic, political, social, environmental, demographic, institutional, and psychic dimensions. Each of these dimensions changes at different rates over time and space. Each dimension also has a dynamic set of attributes. These attributes include, for example, the region's: population; work force; occupational structure, industries, firms and establishments of various kinds and sizes producing a variety of products (raw materials, goods and services); urban and transportation systems; social and economic infrastructures; sets of laws, government policies, incentives and regulations; institutional structures related to education and social and commercial activities; minerals, timber, water, weather, climate and other natural resources or elements; and its location relative to other regions and places.

These endowments of a region can be categorised according to the various attributes of relevance to the particular phenomena under investigation; for example, educational institutions can be classified according to skills, scientific or technical training or experiences they impart. Banking institutions may be classified as to whether or not they provide 'venture' capital; and the region's economic climate may be categorised according to its suitability for high technology firms already established or considering location in the region.

The dynamic attributes related to the various dimensions of a specific region may well be insufficient sources of influence in the region's high technology firms. This would be expected to be the case if the designated region was a metropolitan area, county or state. Indeed, one would expect that the major investment decisions which bring about long term changes in the location and activities of an innovative high technology firm are primarily influenced by forces originating from within the firm and from national and global spatial contexts.

Firms, for example, may be envisaged as nesting in dynamic sets of spatial contexts. These will tend to have different scalar and other kinds of dimensions. The dynamic sets of attributes, associated with the dimensions of the different attributes, associated with the dimensions of the different spatial contexts will generate influences that play important roles in the process of microeconomic industrial change. The task of identifying the spatial and other characteristics of these dynamic spatial frameworks associated with major investment decision making carried out by the firm's key decision makers, is clearly a difficult task. We geographers cannot afford to ignore the challenge of this task, however, because to do so would be to deny our fundamental belief that spatial dimensions play a significant role in explaining the process of industrial change.

Theorising about the explanatory roles of place and spatial connections and context in the process of industrial change has not been impressive at the macroeconomic level (Gore 1984). Unfortunately it is even far less impressive at the microeconomic level of the firm. At this time therefore, there is merit in outlining a set of heuristics to provide useful guidance in seeking a greater understanding of this explanatory role. Discussion is, however, limited to one set of brief comments.

The 'Regional Factor' in a Designated Region

This first set focusses on the search for the dynamic interdependent relationships between a designated region ('the regional factor') and the innovative high technology firms in the region ('the innovation factor'). Causal sequences and intentionality within these relationships between the region and the firms are believed to provide important parts of the mechanism which explains such events as the initial location of the high technology firms in the region, as well as subsequent changes in their size and output characteristics.

One important source of insights is the literature on entrepreneurs and entrepreneurial activities. Over the last fifteen years there has been great interest shown in research on intense concentrations of entrepreneurial activity such as those found in the Silicon Valley, California; Austin, Texas and along Route 128 around Boston, Massachusetts (Bollinger, Hope and Utterback, 1983).

Research on these spatial concentrations, however, has been intuitive and inductive and lacks the guidance of a sound theoretical framework. Current knowledge of the role of the regional factor's influence on entrepreneurial activity is primarily based on case histories and anecdotal evidence and there is a great lack of empirical verification. The literature is, however, rich in insights concerning the relationships between a region's human and non-human endowments and attributes and entrepreneurial activities (Bruns and Tyebjee 1982; Cooper and Komives 1972; Meyer 1978; Vesper and Albaum 1979).

Intriguingly similar spatial relations between a region and entrepreneurial activities, in the form of small high technology firm foundation, have been noted in the literature. Harris

(1984), for example, observes that in Northeastern metropolitan areas in the U.S. entrepreneurs have a great propensity to locate their new firm in or near their hometown. Studies in Silicon Valley (Cooper and Komives 1972) also suggest that technical entrepreneurs tend to start firms in the same geographical area as they are working and living. Environmental information for the 'local region' and information and experience, obtained in an incubator organisation or previous 'start-up' ventures, appear to be important types of information. The prospects of founder entrepreneurs establishing successful 'spin-offs' are affected significantly by the nature and quality of these kinds of information (Bruns and Tyebjee 1982).

Dynamic Role of Regional Factor: Firm's Perspective

From a study of contemporary literature on entrepreneurial activity, innovation and economic development, one may derive expectational propositions, for example: One would expect that the set of critical endowments and attributes of the region, perceived to be very important to an individual high technology firm, will be a different set in small indigenous plants as compared to the set for a similar branch of a non-indigenous firm. For example, the availability of appropriate venture capital in the region would be expected to be far more important for the indigenous establishment than for the branch plant of a non-indigenous firm. The latter firm might well use its own investment resources generated within the firm or obtained from well established sources from outside the region.

New firms in radically different product industries may be expected to have different sets of critical regional environmental factors (endowments and attributes) that have a major influence on their initial location decision. This will tend to be the case also with respect to subsequent location decisions such as deciding whether or not to move elsewhere, stay at the same location, or to close down. This expectation is based on the fact that firms in different industries produce products that require different kinds and/or combinations of inputs from different locations and generally serve different markets. The region's role as supplier of various kinds of inputs and its role as a market for the outputs would be expected to vary more in establishments of similar size across industries than for similar establishments in the same industry.

One expects also that for the same establishment its perception of the influence or significance of specific regional environmental factors, with respect to location decisions or decisions related to its growth, will tend to change over the life of its product(s). The single plant firm's perception will also tend to change in response to organisational changes as it grows and especially if it becomes multiplant and multiregional and multinational in character (Thomas 1975).

The explanatory role of the 'regional factor' with respect to spatial, scalar and structural changes in innovative high technology firms in a region needs to be considered in a temporal framework, because the nature and influence of the region's endowments and attributes change over time and geographic space. Consider, for example, the growing adverse impact on the high technology firms in the Silicon Valley of the markedly deteriorating 'living' environment. Firm dispersion from the region is accelerating: alternate locations are sought by entrepreneurs and firms that would have chosen to locate in Silicon Valley if it had remained as it was a decade or so ago.

Over time changes, for example, in the spatial economic, organisational, technological and market characteristics of the region's high technology firms will also have a dynamic impact on the nature and influence of the regional factor. This appears to be an interdependent, interactive dynamic relationship between the region and its specific set of high technology and non high technology firms.

The Processes of Innovation and Change in Innovative Firms

In this section I will examine briefly the concept of innovation which is defined as "almost any nontrivial change in product or process, if there has been no prior experience" (Nelson and Winter 1977, 48). I will discuss innovation as an element in the complex phenomenon of technical change and as a key element in the technological decision environment of an innovative firm. For these purposes I will use the concepts innovation, technology, product and product industry life cycles as organising frameworks.

The Innovative Process

The innovation life cycle may be conceived as having two components: the first stage is the precommercial which culminates in the commercial production of a technologically 'radically new' product; and the second is the commercial stage.

Only a relatively small percentage of high technology firms are thought to have innovative activities which cover the full spectrum over the innovation life cycle for a radical product. The pre-commercial stage of this cycle begins: - when an invention or set of inventions establish the engineering feasibility of producing a radical product; - a non-routine decision is made by the firm's major decision makers to carry out innovative activities which hopefully will establish an acceptable level of economic feasibility to permit them to produce commercially the radical product.

Innovations which result from carrying out further innovative activities on the 'radically new' product are viewed as incremental in nature. These incremental product and process innovations are meant to improve the radical product and also increase the economic efficiency of its production. The emphasis on the latter type of innovative activities tends to occur when production methods for producing the firm's product are standardised and product enhancement innovative activities become uneconomic.

The Innovation Life Cycle

The simple model of the innovation life cycle thus creates two categories of innovations, the radical and the incremental. A 'radical' or 'epochal' innovation may be viewed as destroying the usefulness of existing competence in key factors of production such as capital, labour components, management and organisation. For example, the introduction of closed steel bodies in the 1920s by the Ford Motor Company meant the demise of the Ford Model T. The decision to use closed steel bodies, among other effects, resulted in the replacement of 15,000 machine tools, the introduction of new processes and the laying off and the hiring of thousands of workers in the giant, efficient integrated plant that Ford created in Detroit to produce the Model T (Clark 1983).

Unfortunately, we do not at present possess a metric for deciding clearly and accurately how 'radical' a particular innovation may be. It seems clear, however, that the commercial production of a radical product innovation signals a discontinuous change in the path of technical progress and a future for the product that is fraught with both technical and commercial uncertainties (Clark 1983).

Radical product innovations represent the Schumpeterian type of innovation that provide the innovator with a 'behavioural competitive edge' and generate spontaneous and discontinuous changes in the economy (Schumpeter 1939, 1950). A modified and refined Schumpeterian statement concerning the impact of radical or 'basic' innovations in the economy is provided by Freeman, Clark and Soete (1982).

Incremental innovations represent those product and process innovations which are often thought of as sources of continuing technical change or 'normal technical progress' (Dosi 1984, 90). The addition of more powerful and more fuel efficient engines to the Boeing 737 commercial aircraft represented incremental innovations.

In our microeconomic behavioural theories of industrial change both radical and incremental innovations are accorded important explanatory roles (Nelson and Winter 1982). Nevertheless many of the processes associated with these roles do need to be more clearly specified and understood. I propose therefore, to examine a number of concepts and conceptual frameworks which appear to be useful in the search for ways and means of strengthening these aspects of this body of theory.

Definitions

At this time there is merit in defining key concepts. Technology
is defined as:

> a set of pieces of knowledge, both directly 'practical'...
> and 'theoretical',...know how, methods, procedures,
> experience of successes and failures and also, of course,
> physical devices and equipment (Dosi 1984, 83).

When exploring the patterns of change in technology there is
utility in using the concept technological paradigm (Dosi 1984,
83). Similar concepts include 'technological regime'(Nelson and
Winter 1982), 'technological imperatives' (Rosenberg 1976),
'dominant design' (Abernathy and Utterback 1978) and
'technological guide-posts' (Sahal 1982). In a broad
impressionistic analogy with the Kuhnian definition of a
scientific paradigm, a technological paradigm is defined as: "a
model and a pattern of solution of selected technological
problems, based on selected principles derived from natural
sciences and on selected material technologies" (Dosi 1984, 83).

One must be careful, however, not to present any suggestion
that there is in existence a unidirection deterministic sequence
of 'science-technology-production'. The nature and direction of
influence of the complex relationships among science, technology
and production represent topics subject to considerable
contemporary debate (Freeman 1979, Price 1984, Walsh 1984).

Within specific sets of technologies associated with
specific technological paradigms there seems to exist positive
and negative heuristics which embody strong prescriptions on the
directions of technical change to pursue and those to neglect
(Dosi 1984, Abernathy 1978). Innovative activities are assumed to
be purposeful and the heuristic search process within a paradigm
involves the use of a set of procedures for identifying,
screening and homing in on promising ways to get to, or close to,
an objective. The procedures may be characterised by the use of
prominent targets, certain cues and clues, and various rules of
thumb (Nelson and Winter 1977, 52-53).

The use of a heuristic or even a stable widely used set of
heuristics or 'heuristic strategy' does not guarantee neither
desirable nor unique outcomes. There is a great deal of trial and
error and emphasis on learning processes involved in product
innovation. Clearly the characteristics of the procedures used in
the innovation process underscore the state of technological
uncertainty associated with product innovation.

It would also appear that dealing with this uncertainty is
far from being costless. One would expect these costs to be
proportionately greater the newer the technologies used and the
more technologically radical and complex the product. For the
firm the product innovation R&D costs per unit of output would
probably decline over the innovation life cycle. Dosi (1984, 88)

suggests that:

> the emergence of new technological paradigms is contextual
> to the explicit emergence of economically defined 'needs'.
> In other words, the supply side determines the 'universe' of
> possible modalities through which generic 'needs' or
> productive requirements (which as such do not have any
> direct economic significance) are satisfied.

The technology of the specific technological paradigm may, for
example, be related to generic 'needs' or 'tasks' such as
transporting people and commodities; and switching and amplifying
electrical signals. The latter generic 'need' may be related to
the semi-conductor set of technologies. Nelson and Winter (1977),
Dosi (1984), Sahal (1982) and other researchers have offered
empirical support for the key characteristics of the notion of
the technological paradigm. Nevertheless, further development,
refining and testing of the concept are necessary.

The emergence of technological paradigms or guide-posts
tends to depend on a new synthesis of a great many proven
established concepts or ideas. This creative symbiosis however,
materialises infrequently but when it does, the consequences are
striking and they represent a break with the past in an otherwise
evolutionary system of technical progress. Another characteristic
that serves to identify a technological paradigm is the presence
of a significant technique with greater adaptability to its task
environment. Such a technique would be expected to be the means
for further advances and technical progress (Sahal 1982, 309-
310).

It seems that the intrinsic nature of a technique plays an
extremely important role in determining the effectiveness of the
research carried out. Furthermore, it appears that the roles of
learning and scaling processes are of paramount importance as
determinants of the course of technical progress. Their roles in
the process however, are conditioned significantly by particular
physical properties of the specific technological system such as
its form and structural complexity. In addition, scaling usually
preceeds learning in the microrealm of innovative activity
associated with the establishment of a new technological
paradigm, whereas learning tends to come first and scaling
follows in the macrorealm of incremental innovation after
commercial production of the radical product has been initiated
(Sahal 1982, 308, 313).

Selected problems associated with selected technologies are
usually discussed in relation to a set or cluster of technologies
such as semi-conductor, bio-technical and organic chemistry
technologies. Normal technological progress within a specific
technological paradigm, for example, the semi-conductor
technological paradigm, may be defined by its technological
trajectory. In other words, a pattern of 'normal' problem-solving
activity (i.e. of 'progress') within a technological paradigm

defines a technological trajectory (Dosi 1984, 83). The concepts of 'natural trajectory' (Nelson and Winter 1982), 'focussing device' (Rosenberg 1982) and 'technological trajectory' are used in a similar way.

The technological trajectory is defined by economic as well as technological sets of variables. The trajectory may be thought of as:

> the movement of multi-dimensional tradeoffs among technological variables that the paradigm defines as relevant. Progress can be defined as the improvement of these tradeoffs. One could imagine the trajectory as a 'cylinder' in the multidimensional space (Dosi 1984, 85).

In addition to special elements associated with a specific technology in any paradigm there are two 'natural trajectories' that are common to a wide range of technologies and are well identified in the literature: progressive exploitation of latent scale economies; and increasing mechanisation of operations that have been done by hand (Nelson and Winter 1977, 58).

One can think of very general as well as very circumscribed trajectories, such as respectively that for nuclear power-generation equipment and that for a specific firm producing this kind of equipment. The more powerful the trajectory the bigger the set of technologies which it excludes. It seems that it is also difficult to switch from a powerful trajectory to an alternative one.

A change in paradigm seems to require one to start (almost) from the beginning in the problem solving activity. Know-how transferability is not smooth when there is a change in the problems needing to be solved. It would seem doubtful that one could compare and assess ex ante the superiority of one technological path over another. Uncertainties related to the technological development path and to its economic environment over time would render the results of such an exercise as dubious (Dosi 1984, 85-86). Before there is a committment to a trajectory, technical know-how may have the pliancy of putty. Looking back over the life of a trajectory however, the technical know-how associated with it seems to have the fixity of baked clay. Engineers, like most other human beings, are not known for their flexibility!

Radical and Incremental Innovations in the High Technology Firm

How do we use the concepts of technology, technological paradigm and technological trajectory in improving our conceptualisation of the processes of both radical and incremental innovation and the changes they effect in high technology firms? In these 'science' and/or 'technologically based' firms the search for radical and incremental innovations, in general, will not be a random process. The decision maker(s), or appropriate key advisor, 'seeking' a radical new product is, or are, assumed to have a sound knowledge of a particular paradigm or technological regime related to a specific cluster of technologies, for example, reciprocating engines in aircraft, vacuum tubes and mechanical calculators. As we know, radical or 'revolutionary' (Abernathy and Clark 1985, 12) innovations such as jet engines, solid state electronics and electronic computers, through revolutionary design, have largely overthrown these older technologies and in the process, rendered much technical and production competence obsolete.

A 'revolutionary' design ushers in a new technological paradigm and a new radical product. Design here represents a revolutionary 'adaptation of means to some preconceived end' (Layton 1972). The beginning of the new technological paradigm represents then, the end of the precommercial stage of the radical innovation and the beginning of the commercial stage of the radical product innovation's life cycle. It also represents conceptually, the commercial beginning of what is sometimes referred to as the product life cycle. Furthermore, it may represent the beginning of a conceptually broader but 'fuzzy' industry life cycle made up of a number of closely related individual product life cycles.

Some clarification has been provided, however, by Saviotti and Metcalfe (1984, 141) when they recently developed a conceptual framework that seems potentially useful for defining the characteristics of a radical product. In this framework:

> a product is considered a combination of three sets of characteristics, one describing the technical features of the product, one describing the services performed by the product, and one describing the methods of its production. These sets of characteristics are related by patterns of mapping.

Every product is produced in a firm by means of a process. Every product embodies a particular technology. This product is simultaneously both the output of the firm that produces it and the input of the firm that purchases it. The product is purchased for the services it provides. In these integral dimensions of the product we can visualise dynamic sets of technological and economic interrelationships and interdependences. These are important elements in the decision environment of the firm.

At the point in time when the radical product innovation is produced commercially, its product has a unique set of technical, process and service characteristics. Subsequently, the firm may be able to change the characteristics of the product by changing the technical and process technologies and changing the range and qualities of the services of the resulting products. These changes in all three sets of characteristics represent incremental product innovations. If the changes represent upward movements on the firm's technological trajectory the incremental product innovations represent 'normal' technological progress within the technological paradigm.

Private enterprise firms tend to attempt to profit as long as possible by keeping confidential, or protecting, their economic control over firm specific and plant and product system-specific technology; as well as their organisational and economic expertise. These kinds of information possessed by a firm play a key role in its competitive behaviour and in determining the degree of success it attains in selling its product. The firm's plant and product system-specific information refers to the information possessed by a firm that differentiates each firm from its rivals, and gives a firm its competitive edge. Firm specific information, however, cannot be attributed to any specific product because it results from the firm's overall activities. This classification of the firm's information represents an extension of a classification system of technical information developed by Hall and Johnson (1970). Clearly a firm's information on its product and/or process innovations is incorporated in its plant systems-specific information category. Technical know-how is to a considerable extent both product and plant specific and lacks the ubiquitous relevance of pure science knowledge. Successful technology transfer so often hinges upon very precise alternations in the design of the specific technique being transferred so that it will suit the requirements of differing production systems. Consequently, the transfusion of technical know-how across plant and firm boundaries is invariably time-consuming and costly.

The radical product innovator firm in the early commercial production phase may sell information about its innovation to other firms and these may produce it under license. Other firms may obtain the information through industrial spying; hiring key members from lead innovative firms; and still others develop similar products on their own. We may then conceive of these firms as belonging to and functioning in a common industry within the same technological paradigm or regime. Firms in this industry would be expected to produce products characterised by somewhat different sets of product characteristics at a point in time, as well as over time. These product differences manifest the process of incremental innovation and the intra-industry diffusion of generally 'normal' technological progress. At some point in time in the evolution of the product and its industry, a dominant design for the product is established and unless a revolutionary change in the technological regime of the product occurs, incremental product and process innovation will come to an end (Utterback and Abernathy 1975). In other words:

the rate of return on capital invested in exploiting a particular technological regime will depend upon the price of inputs specified by the process characteristics and upon the user valuations placed upon the various service characteristics of the associated product. This rate of return will be a key determinant of the growth of productive capacity and thus of the penetration of a regime into the economic environment (Saviotti and Metcalf 1984, 144).

When the returns on incremental product and process innovations are not worth the investment the radical product's innovation life cycle comes to a close. Individual firms will reach such a decision at somewhat different times.

Rejuvenation of the Technology Life Cycle Technologies

A rejuvenation in the product's incremental innovation activities may occur, however, if a major change is effected in the process or product technologies. This rejuvenation or 'de-maturity' (Clark 1983, 112-115) of the technology cycle may well occur in a particularly robust technological paradigm where possible technological directions, whose outer boundaries are defined by the paradigm itself, are discovered and prove to be productive. These sources of 'normal' problem-solving activity then will generate a new burst of incremental innovation and renewed normal technological progress may result (Dosi 1984). Clark (1983) suggests that the process of 'de-maturity' in the American automobile industry began in the early 1970s.

The end of the radical product's innovation life cycle, however, usually does not mean the end of the commercial product or industry life cycle. Non technical sources of increases in economic efficiency such as: realisation, through learning processes, of greater efficiencies associated with existing levels of mechanisation and automation in the firm: economies of growth and plant rationalisation; together with a continuation of effective demand for the product permits some firms in the product industry to continue in production after all incremental product innovations have ceased.

Additional Macro and Microeconomic Perspectives

Macroeconomic interindustry aspects

Now the utility of using additional perspectives in conceptualising the innovation life cycle can be illustrated. The use of these perspectives provide a variety of insights concerning the process of industrial change at the microeconomic level. Consider an examination of the process, technical and service characteristics of a product innovation in both its pre-commercial and commercial life cycle stages from the perspective of not only the firms in the innovation industry concerned but also from the perspective of innovative firms in other industries. These other industries are these that usually supply machinery, etc., used to make the innovation product; and the material and/or intermediate product input that define the technical characteristics of the innovation product.

A limited number of studies in which innovation products were examined from these perspectives suggest the following speculative yet seemingly plausible propositions.

Process and product input sourcing firms and product user firms play significant roles in both events leading to the birth of a radical product innovation and subsequently in the process of incremental product and process innovation (Von Hippel 1976). It seems, therefore, that there are important sets of dynamic technological interdependence linkages that exist between innovative firms in different industries and locations. These linkages influence not only the technological nature of the processes used in manufacturing innovation products but also the technological nature of, and services rendered, by the innovation products. In addition they facilitate the transmission into the economy what are thought to be very important input and output quality multiplier effects (Thomas 1969). Furthermore, productivity enhancing impacts are transmitted along these intra-en inter-regional and intra- and inter-industry input-output flow networks (Freeman 1982, Nadiri and Bitros 1978, Mansfield 1981, Link and Long 1981).

Despite severe problems associated with their specification and application, the innovation and product life cycle concepts have provided useful frameworks for studying and conceptualising many changing relationships related to the process of industrial change. There has been, for example, increasing contemporary interest in the venerable question concerning the rate and direction of technical change, that is, the rate and direction of capital/labour substitution.

At a macroeconomic level, Freeman, Clark and Soete (1982) have used the innovation and product life cycle effectively and plausibly in their study of the role of technological innovation in different stages of the current fourth and earlier third Kondratiev long waves. They are especially interested in the employment and unemployment generating impacts of the innovation process. It appears that the continuing substitution of capital for labour is initiated early in the life cycles of product and process innovations and commercial product industries. This kind of substitution is facilitated as process and product technologies offer opportunities for standardisation as the commercial product moves along its life cycle. Capital cost and revenue change implications of the substitution, management-labour relations, and profitability considerations are some of the important factors that influence the rate and extent of this substitution process.

The employment generating impact of innovative activities during the expansionary stages of the Kondratiev long wave tends to offset their unemployment generating impact. In the post World

War II era this pattern of change was especially evident in capitalist economies until the late 1960s. Major factors accounting for this change were the direct and indirect positive income, employment, and technological input-output multiplier effects, associated with the rapid expansion of firms in 'rejuvenated mature' and 'emerging new' high technology industries.

The complex forces which bring about the recession and depression stages of the Kondratiev wave, however, swamp even the expansionary innovative forces found in most contemporary high technology industries. Negative multipliers during these stages take their seemingly inevitable erosive toll on the well-being of national economies, industries, firms and individuals.

The social implications of the changing relationships between capital and labour are of interest to many scholars. They may be usefully studied in Kondratiev long wave and product and industry life cycle frameworks. Little work appears to be underway, however, on these relationships at the level of the firm. Noteworthy also is a growing contemporary research on the process of 'de-industrialisation' and 'structuration' and its impacts on labour and the spatial change tendencies in various industries undergoing major structural change (Bluestone and Harrison 1982, Massey 1984, Castells 1985, Storper 1985, Walker 1985).

The innovation process and microeconomic decision making behaviour

During the last decade and a half, behavioural theories of the firm have been an important source of ideas and guidance concerning the processes of industrial change (Nelson and Winter 1982). One of the critical components of this body of theory is that concerned with the decision making behaviour of the firm. Decision behaviour related to the innovation process is a significant integral part of this area of study (Thomas 1986). The changes in the characteristics of the technological, organisational and economic decision environments of the firm, and the changes in their relationships to each other at different points along the innovation life cycle, appear to have explanatory connections to non-routine decisions. These decisions subsequently are manifested in changes in the firm's capacity, product and spatial tendencies.

Nelson and Winter (1977, 1982) have made seminal contributions to our understanding of the process of innovation and to the development of conceptual tools for the study of why and how major decision makers behave as they do in innovative firms. Their conceptual framework for understanding firm decision behaviour, in these kinds of firms, is developed on the foundation provided by three interrelated notions: decision environments, organisational routine and search.

There appears to be high degree of agreement in the literature that innovative firms operate under varying conditions of uncertainty in their internal and external decision environments. There is, however, still a great need for conceptualisations that increase our understanding as to why and how key decision makers behave when they encounter conditions of uncertainty such as in the process of innovation and in other non-routine capital investment decision making.

Is this behaviour reflective of decisions and actions based on the pure 'unclearned insight' of a Schumpeterian innovator? Or is it behaviour that represents decision making based on 'learned acts of skill' of the top level managers of the innovative firms? Is this decision behaviour different in single and multi-plant and multiregional firms or in firms of different size in the same and in different industries? In what way do variations in the organisational structure of innovative firms influence their decision behaviour (Ettlie and Bridges 1982)?

There is reason to believe that many types of innovations involve little uncertainty (Freeman 1982). Such innovations may involve more managerial 'learned acts of skill' than entrepreneurial 'unlearned acts of insight'. I suggest that we need to address the questions "Do different acts of non-routine decision-making require different proportions of 'unlearned acts of insight' and 'learned acts of skills', and do these proportions vary in specific decisions made over the life time of an innovation product?"

As discussed earlier, in the case of a 'radically new' product, the innovation decision making process extends from the beginning of the precommercial stage through to the end of the commercial stage of the life cycle. Different conditions of uncertainty, in the technology, organisation, economic and market decision environments of a firm probably vary not only over the innovation and commercial life cycles of its products but over the firm's life cycle as well. It seems reasonable to assume that a firm's set of decision responses to these various combinations of different kinds of uncertainty will be reflected in the levels of commercial success achieved, over time, by the firm. It is also reasonable to posit that the tendencies for spatial change in firms will change over the innovation and commercial life cycles of their products. Actual change in their locations will, however, depend on a number of factors such as the innovation cycle characteristics of the product, capacity requirements, production costs, length of product cycle and supply sourcing locations and market dimensions (Thomas 1985). Tentative support for the speculation seems to be provided in a statement by Clark (1983, 105) based on extensive research on the American automotive industry:

> where products and processes are standardized and change only incrementally, technique and know-how tend to be embodied in capital equipment that can be readily purchased

37

for use in many parts of the world. Location and indeed competition then depends much less on technology, and much more on cost.

Information

Information is a counter to uncertainty in the firm's decision environments. The search for information that will help to reduce uncertainty is especially important to innovative firms that wish to be commercially successful. R&D activities represent a major component of the search mechanism of science and technology based firms engaged in innovation. Yet, as mentioned earlier, we know relatively little concerning these activities and the long term R&D investment patterns of innovative firms (Mayer-Krahmer 1984).

Contemporary received theory of the entrepreneur stresses his coordinator role in information search processes within the firm (Casson 1982; Kirzner 1983; Williamson 1983). The theory of heuristic search has provided useful guidance in studying the process of information search and choice of search strategy by firms (Thomas 1986). With respect to the innovative firm we need to know, for example, the nature of the information search processes and how these are conducted with respect to the external and internal environments of the firm; how is the information transmitted to the decision-makers; how is information degradation handled in the firm; how is the information search and its implementation organised and coordinated in different types of firms as well as when decision-makers in the firm are dealing with different types of non-routine decisions such as location decisions. Schmenner (1982) provides an insightful account of the process of location decision making in a variety of firms. He notes the use by the firm of sets of heuristics in this process. Of course, we also need to know the 'why?' as well as the 'how?' with respect to these behaviours and activities if they are to be explained.

Concluding Statement

In this chapter an explanatory discussion of a number of ideas which connect and relate conceptualisations of innovators, innovations and innovative firms to the process of industrial change in capitalist economies is initiated. An important foundation for the discussion was provided by the generally accepted perceptions that
1. innovation provides the cutting edge for advancing technology;
2. advancing technology is a major permissive source of economic growth and change.

Firms as organisational entities represent useful but underutilised vehicles for studying the process of innovation and industrial change at the microeconomic level. Commercial success or lack of success of a firm's innovative activities are reflected in changes in its capacity, product and spatial dimensions.

The chapter was concerned with the development of a way of conceptualising the process of innovation and change in dynamic, science and technology based innovative firms within a spatial context. The first dealt with the conceptualisation of a variety of dynamic spatial and temporal environments which constitute the spatio-temporal contexts within which the innovative firms operate. These environments generate their own sets of influences which impact the firm. Therefore they need to be considered explicitly as parts of the dynamic decision environment of the firm. Thus, there is also a need to discover the various heuristic search mechanisms used by the innovative firms when they carry out their assessment of various dimensions of their spatio-temporal contexts.

In the second section, it was shown that a number of related concepts such as the radical product innovation and commercial product life cycles, technological paradigm, technological trajectory and innovation product characteristics have considerable potential value in our search for a better understanding of the technological dimensions of the process of innovation. They also provide useful insights concerning the role of innovation in the process of industrial change at macro- and microeconomic levels.

Additional sets of related concepts were discussed in the third section and these were focussed on the relationships between selected decision environments and decision behaviour within innovative firms. It was shown that innovative firms operate under varying conditions of uncertainty in their decision environments. To offset these conditions commercially successful firms tend to develop effective ways of both searching for, and using appropriate information to facilitate the attainment of an acceptable measure of their goals and objectives. Nelson and Winter's framework, based on the three interrelated notions of decision environments, organisational routine and search, has proven to be a productive paradigm for studying the decision behaviour of innovative firms. The role of various sets of heuristics in different components of the decision behaviour of a firm represents a research topic of major importance.

Clearly the conceptualisations presented need considerable refinement, development and testing. Nevertheless, they do provide helpful insights and information concerning the process of innovation in innovative, science and technology based firms. They also show how innovation plays a major role in effecting micro and macroeconomic long term industrial changes at various geographical scales. The task of providing ways of achieving a better understanding of why, how and where innovation occurs in firms, industries and geographic space remains an important and worthwhile challenge.

Note

I wish to thank Günter Krumme and J. Scott MacCready for their helpful comments. This material is based partly on work supported by the National Science Foundation under Grant No. SES-8411682.

References

Abernathy, W.J. (1978) The Productivity Dilemma, Johns Hopkins, Baltimore

Abernathy, W.J. and K.B. Clark (1985) 'Innovation: Mapping the winds of creative destruction' Research Policy, 14, 3-22

Abernathy, W.J. and J.M. Utterback (1978) 'Patterns of industrial innovation' Technology Review, 80, 40-47

Armington, C., C. Harris, M. Odle (1983) Formation and growth in high technology businesses: a regional assessment (mimeo), The Brookings Institution, Washington, D.C.

Bluestone, B. and B. Harrison (1982) The Deindustrialization of America, Basic Books, New York

Bollinger, L., K Hope and J.M. Utterback (1983) 'A review of literature and hypotheses on new technology-based firms', Research Policy, 12, 1-14

Brugman, B.L. (1983) A Spatial Perspective on the Process of Technological Innovation in Technology-Intensive Industry, Unpublished Ph. D. dissertation, University of Washington

Bruns, A.V. and T.T. Tyebjee (1982) 'The environment for entrepreneurship' In C.A. Kent et al. (Eds.) The Encyclopedia of Entrepreneurship. Prentice-Hall, Englewood Cliffs, N.J. Chapter XVI, 288-315

Casson, M. (1982) The Entrepreneur: An Economic Theory, Martin Robertson, Oxford

Castells, M. (1985) 'High technology, economic restructuring, and the urban-regional process in the United States' In M. Castells (ed.) High Technology, space and society, Sage Publications, Beverly Hills

Clark, J., C. Freeman and L. Soete (1984) 'Long waves, inventions, and innovations' In C. Freeman (ed.) Long Waves in the World Economy, Frances Pinter, London

Clark, K.B. (1983) 'Competition, technical diversity, and radical innovation' Research on Technological innovation, management and Policy, 1, 103-149.

Cooper, A.C. and J.L. Komives (Eds.) (1972) Technical Entrepreneurship: A Symposium, Center for Venture Management, Milwaukee

Dosi, G. (1984) 'Technological paradigms and technological trajectories. The determinants and directions of technical change and the transformation of the economy' In C. Freeman (ed.) Long Waves in the World Economy, Frances Pinter, London

Ettlie, J.E. and W.P. Bridges (1982) 'Environmental uncertainty and organizational technology policy' IEEE Transactions on Engineering Mangement EM-29, 2-10

Freeman, C. (1979) 'The determinants of innovation' Futures, 11, 205-215

Freeman, C. (1982) The Economics of Industrial Innovation, MIT Press, Cambridge Mass.

Freeman, C., J. Clark and L. Soete (1982) Unemployment and Technical Innovation, Westport, Greenwood Press, Connecticut

Glasmeier, A.K., P.G. Hall and A.R. Markusen (1983) Recent evidence on high-technology industries' spatial tendencies: a preliminary investigation, Working Paper No. 417. Institute of Urban and Regional Development, University of California, Berkeley

Glasmeier, A.K. (1985) 'Innovative manufacturing industries: spatial incidence in the United States' In M. Castells (ed) High Technology, Space and Society, Sage Publications, Beverly Hills

Gore, C. (1984) Regions in Question: Space, Development Theory and Regional Policy, Methuen, London.

Hall, G.P. and R.E. Johnson (1970) 'Transfers of United States aerospace technology to Japan' In R. Vernon (ed.) The Technological Factor in International Trade, Columbia University Press, New York

Hall, P. and A. Markusen (Eds.) (1985) Silicon Landscapes, Allen and Unwin, Boston

Harrington, J.W. (1983) Locational Change in the U.S. Semiconductor Industry, Unpublished PH.D. dissertation, University of Washington

Harrington, J.W. (1985) 'Corporate strategy, business strategy, and activity location' Geoforum, 16 4, 349-357

Harrington, J.W. (1985) 'Intra industry structural change: U.S. semiconductor manufacturing' Regional Studies, 19, 343-352

Harris, C. (1984) High-technology entrepreneurship in metropolitan areas, Brookings Institution (unpublished), Washington, D.C.

Jones, F. and R. Struyk (1975) Intra-metropolitan Industrial Location: The Pattern and Process of Change, Lexington Books, Lexington

Kirzner, I.M. (1983) 'Entrepreneurs and the entrepreneurial function: a commentary' In J. Ronen (ed.) Entrepreneurship, Lexington Books, Lexington, Mass.

Layton, E. (1972) 'Technology as knowledge' Technology and Culture, 15, 31-41

Link, A.N. and J.E. Long (1981) 'The simple economics of basic scientific research: a test of Nelson's diversification hypothesis' The Journal of Industrial Economics, 30, 105-109

Malecki, E.J. (1979) 'Locational trends in R&D by large U.S. corporations, 1965-1977' Economic Geography, 55, 309-323

Malecki, E.J. (1980) 'Dimensions of R&D location in the United States' Research Policy, 9, 2-22

Malecki, E.J. (1980) 'Firm size, location and industrial R&D: disaggregated analysis' Review of Business and Economic Research, 16 29-42

Malecki, E.J. (1980) 'Corporate organization of R&D and the location of technological activities' Regional Studies, 14, 219-234

Malecki, E.J. (1981) 'Science, technology and regional development: reviews and prospects' Research Policy, 10, 312-334

Malecki, E.J. (1982) 'Federal R&D spending in the United States of America: some impacts on metropolitan economics' Regional Studies, 16, 19-35

Mansfield, E. (1981) 'Composition of R&D expenditures: relationship to size of firm, concentration and innovative output' Review of Economics and Statistics, 63, 610-615

Markusen, A. (1984) Defense spending and the geography of high tech industries, Working paper 423. University of California, Institute of Urban and Regional Development, Berkeley

Markusen, A. (1984) Military spending and Urban development in California: introduction to study report, University of California, Department of City and Regional Planning, Berkeley

Markusen, A. (1984) Defense spending: a successful industrial policy? Working paper 424. University of California, Institute of Urban and Regional Development, Berkeley

Massey, D. (1984) Spatial Divisions of Labour: Social Structures and the Geography of Production, Macmillan, London

Mayer-Krahmer, F. (1984) 'Recent results in measuring innovation output' Research Policy, 13, 175-182

Meyer, J.W. (1978) 'Strategies for further research: varieties of environmental variation'. In M.W. Mayer (ed.) Environments and Organisations, Jossey Bass, San Francisco.

Nadiri, M. and G. Bittros (1978) 'Research and development expenditures and labor productivity at the firm level' In J. Kendrick and B. Vaccara (Eds.) New Developments in Productivity Measurement, National Bureau of Economic Research, New York

Nelson, R.R. and S.G. Winter (1977) 'In search of useful theory of innovation' Research Policy, 6 36-76

Nelson, R.R. and S.G. Winter (1982) An Evolutionary Theory of Economic Change, The Belknap Press of Harvard University Press, Cambridge, Mass.

Oakey, R.P. (1983) Research and Development Cycles, British and American Small High Technology Firms, Discussion Paper No, 48. Centre for Urban and Regional Development Studies, University of Newcastle Upon Tyne

Oakey, R.P.(1984) High Technology Small Firms, Frances Pinter, London

Oakey, R.P. (1985) 'High-technology industry and agglomeration economies' In P. Hall and A. Markussen (Eds.) Silicon Landscapes, Allen and Unwin, Winchester, Mass.

Pavitt, K. (1979) 'Technical innovation and industrial development' Futures, 11 December 458-470

Price, D. (1984) 'The science/technology relationship, the craft of experimental science, and policy for the improvement of high technology innovation' Research Policy, 13, 3-20

Rees, J., R. Briggs and D. Hicks (1985) 'New technology in the United States' machinery industry' In A.T. Thwaites and R.P. Oakey (Eds.) The Regional Economic Impact of Technological Change, Frances Pinter, London

Rosenberg, N. (1976) Perspectives on Technology, Cambridge University Press, Cambridge

Rosenberg, N. (1982) Inside the Black Box: Technology and Economics, Cambridge University Press, Cambridge

Sahal, D.(1982) Patterns of Technological Innovation, Addison-Wesley, Reading, Ma.

Saviotti, P.P. and J.S. Metcalfe (1984) 'A theoretical approach to the construction of technological output indicators' Research Policy, 13, 141-151

Saxenian, A. (1981) Silicon chips and spatial structure: The industrial basis of urbanization in Santa Clara County, California, Working paper No. 345. Institute of Urban and Regional Development, University of California, Berkeley

Saxenian, A. (1985) 'Silicon Valley and Route 128: regional prototypes or historic exceptions?' In M. Castells (ed) High Technology, Space, and Society, Sage Publications, Beverly Hills

Schumpeter, J.A. (1939) Business Cycles, McGraw-Hill, New York

Schumpeter, J.A. (1950) Capitalism, Socialism and Democracy, Harper and Row, New York

Schmenner, R.W. (1982) Making Business Location Decisions, Prentice-Hall, Englewood Cliffs, N.J.

Scott, A.J. (1983a) 'Industrial organization and the logic of intra-metropolitan location: theoretical considerations' Economic Geography, 59, 233-250

Scott, A.J. (1983b) 'Industrial organization and the logic of metropolitan location: a case study of the printed circuits industry in the Greater Los Angeles region' Economic Geography, 59, 343-367

Steed, G.P.F (1982) Threshold Firms, Canadian Government Printing Centre, Hull, Quebec.

Storper, M. (1985) 'Technological and spatial production relations: disequilibrium, interindustry relationships and industrial development' In M. Castells (ed.) High Technology, Space, and Society, Sage Publications, Beverly Hills

Thomas, M.D. (1975) 'Regional economic growth: some conceptual aspects' Land Economics, 45, 43-51

Thomas, M.D. (1975) 'Economic development and selected organizational and spatial perspectives of technological change' Economie Appliquee, 29, 379-400

Thomas, M.D. (1985) 'Regional economic development and the role of innovation and technological change' In A.R. Thwaites and R.P. Oakey (Eds.) The Regional Economic Impact of Technological Change, Frances Pinter, London

Thomas, M.D. (1986b) 'Growth and strutural change: the role of technical innovations' In A. Amin and J. Goddard (Eds.) Technological Change, Industrial Restructuring and Regional Development, George Allen and Unwin, London

Thwaites, A.T., R.P. Oakey and P. Nash (1981) Industrial and Regional Development, Final Report, Centre for Urban and Regional Development Studies, University of Newcastle Upon Tyne

Thwaites, A.T., A. Edwards and D. Gibbs (1982) The Interregional Diffusion of Production Innovations in Great Britain, Final Report, Centre for Urban and Regional Development Studies, University of Newcastle Upon Tyne

Utterback, J.M. and W.J. Abernathy (1975) 'A dynamic model of process and product innovation' Omega, 3, 639-656

Vesper, K.H. and G. Albaum (1979) The role of small business in research, development technological change and innovations in region 10, Working Paper, School of Business Administration, University of Washington

Von Hippel, E. (1976) 'The dominant role of users in the scientific instruments innovation process' Research Policy, 5, 211-234

Walker, R.A. (1985) 'Technological determination and determinism: industrial growth and location' In M. Castells (ed.) High Technology, Space, and Society, Sage Publications, Beverly Hills

Walsh, V. (1984) 'Invention and innovation in the chemical industry: demand-pull or discovery-push?' Research Policy, 13, 211-234

Weintraub, E. (1979) Micro-foundations, Cambridge University Press, Cambridge

Williamson, O.E. (1983) 'Organizational innovation: the transitional approach' In J. Ronen (ed.) Entrepreneurship, Lexington Books, Lexington.

CHAPTER THREE

RESEARCH, DEVELOPMENT AND INNOVATION IN THE SEGMENTED ECONOMY: SPATIAL IMPLICATIONS

Dr. C.S. Morphet, Senior Lecturer in Geography, Newcastle upon Tyne Polytechnic, UK

Introduction

Research on the location of industrial Research and Development (R&D) and technological innovation has been dominated by what can be called a 'supply-side' model. It has been supposed that it is the availability of inputs to the innovation process which determines the innovativeness of industry in any particular region. Thus the pioneering work by Buswell and Lewis (1970) noted the preponderance of R&D establishments in the South East of England and emphasised the importance of labour supply and communications as explanatory factors. The availability of skilled and qualified manpower was associated with the residential preferences of such people.

More recent work by Oakey et al. (1982) argued that regional variations in factors known to be associated with successful innovation - large firm size, the employment of professional technical and market research staff, contact with the wider research community - would suggest that the potential for technological advance would vary between regions.

In a study of three manufacturing MLHs (the smallest industrial grouping (represented by three digits) in the British Standard Industrial Classification) Thwaites (1982) found a pattern of innovation that was largely in accordance with those expectations. In particular it was discovered that while process innovations were not unevenly distributed between British regions (and branch plants located in Development Areas were highly innovative) product innovation was substantially lower in the Development Areas than in the South East, in particular within independent enterprises. Process innovation was defined to include the adoption of existing and proven techniques and, not surprisingly, its occurence was not found to be systematically related to on-site R&D. Product innovation, in contrast, was positively related to research on-site.

The Regional Studies Association (1983, 51) notes these and other similar results and their apparent implications for regional policy. Reference is made to "a regional economic milieu

45

in the South East which is much more conducive to innovation than that likely to be found in the more peripheral regions" and it is suggested that "It seems clear that an increase in the R&D capability of industry in less favoured regions would be a beneficial development". While references to the demand for technological innovation are not entirely absent from the works cited above, the undeniable emphasis is on supply-side considerations. It is generally implied that it is the availability of inputs to the innovation process - R&D, technical information, development capital - that is the determining influence on the volume of technological innovation that the individual firm achieves.

Buswell, Easterbrook and Morphet (1985) have, in contrast, recently emphasised a demand-side view of technological change in a regional context, emphasising the impact of spatial variations in demand in the highly research intensive defence-related industries, and also introducing the idea of variations in innovation strategy, suggesting that these might be associated with differences within the segmented economy (Taylor and Thrift 1983). The purpose of this contribution is to further emphasise the demand-side view of technological change, and to elaborate the postulated connection between regional variations in the segmented economy and regional variations in R&D.

Supply-Side and Demand-Side Views of Technological Innovation

The relative importance of supply-side and demand-side influences on technological innovation has long been contested. Until recently, most economists treated the sources of innovation as exogenous to the economic system, based somewhere in an activity called 'science'. The pioneering works of Schumpeter in the first half of the twentieth century served to secure this presumption. But in the post-war world, and at a time of rapidly rising research budgets and the evident harnessing of science and technology to industry and government, the links between the sources of invention and the economic system came under closer scrutiny. In particular it came to be recognised that a simple linear-sequential view of innovation stemming from science was less than adequate.

Williams (1967) questioned the "...widespread belief that the speed of industrial innovation is a simple function of research" and noted "...how often we are told that we ought to be doing more research in order to ensure a higher rate of innovation (or growth)".

The inadequacy of this supply-side view of innovation had been recognised by Carter and Williams (1957, 108). They wrote:

There can be no scientific progress without scientists, but it is not at all certain that, if more scientists are trained, unprogressive firms will leap into progressiveness.

The various necessary conditions of progress - trained men, money, receptive managements, favourable markets - are essential to each other like the wheels of a clock. But what winds the clock or makes it tick? What are the stimulants of progressiveness? ... Some people suppose that the part of clock-winder is played by the inventor, the man of an original turn of mind, whose flow of new ideas and improvements keeps his firm constantly moving forward. This view is ... out of date because, in the larger firms of today, the flow of new ideas for product and process innovation results, not from the chance inspiration of exceptional individuals, but from a deliberate decision by management to spend money on research and development.

Schmookler (1966) supported such a demand-side interpretation of innovation by showing that innovative activity (as measured by patents registered) correlated with, but lagged behind, investment activity. He argued that invention should be seen not as a variable exogenous to the economic system but as part of that system being directed and sustained in the light of anticipated economic rewards.

A number of other studies conducted during the 1960s further emphasised the demand side of successful technological innovation, introducing the concepts of 'need-pull' and 'market-pull' alongside the supply-side influence of 'technology-push'. Such studies have been reviewed by, for example, Utterback (1974, 621) who writes: "market demand factors appear to be the primary influence on innovation. From 60 to 80 per cent of important innovations in a large number of fields have been in response to market demands and needs..." while Gilpin (1975, 65) puts the point even more strongly:

Everything that we know about technological innovation points to the fact that user or market demand is the primary determinant of successful innovation. What is important is what consumers or producers need or want rather than the availability of technological options. Technological advance may be the necessary condition for technological innovation and on occasion new technology may create its own demand but in general and in the short-run, the sufficient condition for success is the structure or nature of demand

Freeman (1982, 211) represents such views as an extreme interpretation of Schmookler, leading to the belief that:"technical innovation could be regarded as a secondary phenomenon and taken for granted, since it would simply respond to demand management." Freeman offers a criticism in support of that of Mowery and Rosenberg (1979, 104) who question the now widely held assumption that: "the governing influence upon the innovation process is that of market demand; innovations are in some sense 'called' or 'triggered' in response to demands for the satisfaction of certain classes of needs". They criticise the methodologies of a number of the studies that are frequently cited in support of this position, and point out that concepts of demand-side influence have not been used consistently in this body of work.

Perhaps the evident ambiguity in concepts such as 'need-pull' or 'demand' is in part due to the different scales at which such concepts might be applied. Consider a hypothetical situation in which a large research laboratory within some industry develops, as an indirect consequence of its research programme, some process innovation for which it has no use itself. It might set about finding a firm which will purchase or license the design. Subsequently, some potential recipient might be seeking a technical solution to some identified 'need'. If a match occurs and innovation takes place, is this to be seen from the point of view of the industry as 'technology-push' or from the point of view of the recipient firm as 'demand-pull'? To classify it as either would fail to recognise the non-linear nature of the process: the independent identification of market potential and technical potential and the subsequent coupling of these in the act of innovation. Only when the prior existence of one of these recognitions results in the other being deliberately sought and identified, can we meaningfully talk of 'discovery-push' or 'market-pull'. Thus a distinction must be drawn between the dependent and independent components of the coupling process. Where the identification of technology is dependent on the prior (and independent) identification of need or market we have a genuine example of need or market-pull. Where the identification of a market is dependent on the prior (and independent) identification of a technical possibility then we have a genuine example of technology push.

There clearly are such instances where innovation is purely supply-side driven - 'technology-pushed'. Thus any research programme which generates potential applications which were neither sought nor perceived at its outset will, if there proves on subsequent search to be a market for these applications, provide us with an appropriate example. Langrish et al. (1972) studied 84 significant innovations and found that only five could be clearly categorised as 'discovery-push'; they also found that larger technological changes tended to be of this 'discovery-push' type. It might be expected that incremental innovations will tend to be 'market-pull' while 'breakthroughs' will tend to be 'discovery-push' simply because of the difficulty, if not impossibility, of embarking on a demand-led search for a breakthrough. Breakthroughs are by their very nature impossible to specify in advance. Incremental innovations are, in contrast, well suited to prior specification and generation by the means of some planned research programme. Freeman (1982, 211) offers the compatible suggestion that "...once a major innovation has been made, then a pattern of demand-led secondary inventions and innovations may set in over many decades giving apparent credibility to a 'Schmookler'-type of analysis".

It might nevertheless be noted that the type of research which might beget supply-side driven innovation is relatively uncommon. If we use the categories suggested by Rothschild (Cmnd 4814, 1971) there is little evidence that 'basic' research serves to drive innovation (and certainly not for any but the largest corporations) and 'applied' research funded according to the customer-contractor principle is by definition demand oriented. Dainton (Cmnd 1971) identifies this as 'tactical' research.

If we further admit a third category of research condemned by Rothschild but defined by Dainton as 'strategic research' (Cmnd 1971, 3) "the broad spread of more general scientific effort which is needed as the foundation for this tactical science" we might anticipate that this is the type of research which provides a supply-side push to innovation. But even this type of research is rare. Officially (at least) it does not exist in Government research programmes which are either basic or applied (after Cmnd 1971).

Industrial R&D is only rarely classified as basic. Around 2 per cent of current expenditure was thus classified in 1981 (British Business 1983) while 71 per cent of R&D expenditure was classified as development. Official statistics do not allow us to distinguish between strategic and tactical applied research, but the total applied category accounted for only 15 per cent of research and development conducted in private industry (rather more in Research Associations and Public Corporations).

It might be reasonably concluded from the foregoing that most innovative activity will be demand-led, derived from specific R&D programmes organised to achieve specific objectives. There will plainly be exceptions - major innovations arising from longer term strategic research - but these (although often highly significant) will be rare and (as will be argued later) almost certainly confined to a small number of firms. Most firms who do research will operate goal-oriented research programmes resulting in the demand-led secondary innovations to which Freeman (1982) refers.

The Bases of Non-Innovation

We can now turn to those firms which do not undertake Research and Development. It is incidental to our immediate purposes that such firms might be over-represented in peripheral regions. Exhortations to such firms to do more research do not emerge just from regional analysts but can be found directed from a national perspective over a number of decades.

If we are asking that firms should do research we might consider what sort of research we are asking them to do. Nobody would surely suggest that firms which at present do no research should embark on basic or strategic research programmes in order to benefit from technology-push innovations. What is being suggested, then, is that firms should embark on demand-led research programmes.

To suggest this is to make a number of linked assumptions. It is to assume:
1. That demand for technological innovation exists for the firm;
2. That such demand is either
   a. unrecognised;
   b. recognised but not acted upon.

The second assumptions, consequent upon the first, amount to claims of ineffective management. It is assumed that the firm's managers are not doing what they ought to be doing. While this is doubtless true in a number of cases, and Metcalfe (1970) and Mansfield (1971) document instances where an almost stubborn refusal to innovate has led to the demise of firms, there is perhaps just a whiff of arrogance in those academic commentators who purport to know how management should manage. What if it is not the second assumptions that we should concern ourselves with but the first? Might it not be that at the level of a number of individual firms in an industry there is, simply, no demand for the sorts of technological innovations that require an in-house R&D capability?

Variations in the Demand for Innovation

Freeman (1982) offers a vital insight into the innovative behaviour of the individual firm. He points out that rational profit-maximising behaviour is seldom possible in the face of the uncertainties associated with individual innovation projects. Thus firms may adopt certain strategies when confronted with the prospect of technological change. Not all of these strategies are innovative, but even a non-innovative strategy may enable a firm to survive and even prosper. "There are various alternative strategies which they may follow, depending upon their resources, their history, their management attitudes and their luck" (Freeman 1982, 170).

The six strategies which Freeman describes are summarised in Table 1. Freeman is at pains to point out that these are 'ideal types', useful for the purposes of conceptualisation. There is an infinite gradation between the types, and firms may exhibit characteristics of more than one type; they may also change from one strategy to another or follow different strategies in different sectors of their business.

With reference to Table 1 we are now in a position to examine several categories of firm which will have low or non-existent levels of R&D. We should note that in fact most firms fall into this category. In Britain in 1978, the 53 largest firms accounted for 66 per cent of all R&D performed (Business Monitor MO14, 1980) and the majority of small firms (probably over 95 per cent, according to Freeman 1982) do not perform any specialised R&D programmes. Many of these firms will be following

Table 1. Innovation Strategies, after Freeman, 1982

INNOVATION STRATEGY	OFFENSIVE	DEFENSIVE	IMITATIVE	DEPENDENT	TRADITIONAL	OPPORTUNIST
ROLE OF IN-HOUSE SCIENTIFIC & TECHNICAL FUNCTIONS	Aim to produce new products first	Follow, possibly leapfrog, the offensive innovators	Acquire know-how and licenses from offensive & defensive innovators	Innovate on request from (and to specification of) customer	Nature of product requires only design (e.g. fashion) changes	Identify 'niche' in rapidly changing markets
RESEARCH DEVELOPMENT	Highly R&D intensive, bias towards fundamental research	Highly R&D intensive, less emphasis on fundamental research	Lower R&D intensity. Development work predominates	Little or no R&D	None	None
PRODUCTION ENGINEERING	4 Secondary to innovation	4	5 Crucial to viability. Process innovation necessary	5 Crucial-Production oriented	5 Sometimes craft-based	1
SCIENTIFIC EDUCATION & TRAINING	5 Necessary education of customers & own personnel	4 Associated with advertising, selling, technical services	3	3	1	1
LONG TERM FORECASTING & PLANNING	5 Basic identification of future products	4	3	3	1 Reactive	5 Perhaps 'intelligence' rather than planning is the crucial input
BASIS OF VIABILITY OF STRATEGY	High-risk, high-reward. Risks have to be spread	Lower risk	Non-innovation related advantages necessary (markets etc.) Production orientation	Needs low overheads or entrepreneurial skill. Vulnerable to takeover	Vulnerable to exogenous technical change	Imaginative entrepreneurship

'traditional' innovation strategies because they are in traditional industries where the technology can be classified as 'mature' and low research intensities are the norm. The life cycle of a technology has been summarised by among others Rothwell (1982) who points to the increased production orientation in the maturity stage of such a cycle. Within industries with mature technologies (and the clothing industry with a research intensity of less than 0.1 per cent in 1978 affords an example) process innovation will dominate and this will usually be bought in from machinery suppliers. Technical skill may be necessary to effectively incorporate such process innovation into the firm but this may well be provided by the supplier, and even where it is not it is unlikely to merit the title R&D.

In more R&D intensive industries where innovation is the prime means of corporate competition, smaller firms may be unable to meet the financial requirements of the lowest absolute level of R&D which is necessary to support a minimal imitative innovation strategy. Whereas in some research intensive industries (e.g. some types of electronic instruments according to Freeman 1982) this threshold value is low and even firms with a small market share may still be able to support it, in other innovative industries, e.g. communication systems, the threshold will be very high and smaller firms will be unable to survive.

Variations between Firms - Industrial Segmentation

It is crucial to the present argument to note that within any industry all firms are not the same. In particular, all firms do not face the same markets. It is widely understood that the most important customers for manufacturing firms are other manufacturing firms, and networks of material linkages exist within and between manufacturing product groups. Thus within any product group there will be individual organisations of various sizes performing a variety of roles. Some will be large firms producing (at least in part) for final demand, selling to consumers or perhaps to government. Other firms will be acting as subcontractors, or even sub-subcontractors, to them. These subcontracting relationships will extend across into other product groups, where there will be found both customers and suppliers.

Thus across an individual product group there will be a wide variety of determinants of demand. While some firms might enjoy an oligopolistic position in respect of consumer markets, others might face a market composed of government defence contracts, others might face the purchasing departments of large corporations which will place them in a subordinate role. If variations in the structure of demand can be held to account for variations in innovative performance, then there can be little reason to assume a homogenous standard of performance within any industry or product group.

Fig. 1. The Current Pattern of Segmentation in the Segmented
Economy

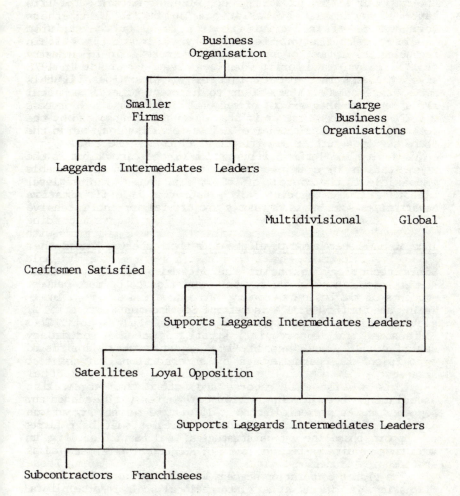

Source: Taylor and Thrift 1983

The profound qualitative differences that exist between industrial firms, and which defy unidimensional categorisation on the basis of either product group, size, or market structure, have been addressed by Taylor and Thrift (1983, 450). The segmentation of industrial firms is to be on the basis of "characteristic economic objectives, labour processes, technological usages, organisational structures, and resources of capital". Taylor and Thrift's business segments are to be seen, like Freeman's innovation strategies, as ideal types. Different parts of the same firm may belong to different segments and whole firms may move into and out of segments over time.

The basic distinction in the segmented economy is between the smaller firm and the large business organisation, but within these segments further significant distinctions are made. Fig. 1 shows Taylor and Thrift's illustration of the current pattern of segmentation in a developed national economy. It will be impossible in the context of the present paper to effectively elaborate the categories which Taylor and Thrift provide. Nevertheless some explanory notes are offered in Table 2.

Innovation Research and Development in the Segmented Economy

Since technology is one of the determinants of industrial segmentation, there is clearly an association to be made between segments of the business economy, and innovation strategy. Indeed Taylor and Thrift (1983) have referred to Freeman's categories in elaborating the ideal types of the segmented economy. Thus 'offensive' and 'opportunist' strategies are introduced to describe the 'leader' small firm segment, while the 'loyal opposition' is characterised by a 'traditional' innovation strategy.

This section will reverse and extend this argument by associating Freeman's innovation categories with each of the segments of the segmented economy. It will be further argued that particular scientific and technical strategies will be required to support these innovation strategies, and particular segments will thus require particular forms of R&D and/or other scientific and technical functions.

The global corporation segment will be excluded from further discussion. Its impact at an international scale is crucial, and its use of technology in particular to decompose the production process (Froebel et al. 1980) is very important. Nevertheless for the present purpose this sector can be included as a special form of the multidivisional segment.

A. The Small Firm Sector
1. The Laggard Segment
   a. Craftsmen. Such firms utilise limited technology. The expected innovation strategy is 'traditional'. There will be no specialised R&D, or scientific and technical

Table 2. The Segmented Economy, after Taylor and Thrift, 1983

A. THE SMALLER FIRM SEGMENT. - may be manufacturing but dominant in retail, service etc.

    1. The Laggard Segment - small independents, high rate of replacement, little growth prospects.
      a. Craftsmen - limited technology for small markets.
      b. Satisfied - proprietor wishes to stay small.

    2. The Intermediate Segment - survival usually linked to large firm sector.
      a. Satellites - either subcontractors to large firms, or franchisees. Subcontracting is common in manufacturing. Large firms can thus externalise risks and take advantage of lower wages paid.
      b. Loyal Opposition - occupying market niches (product or location) which are of little interest to large firms.

    3. The Leader Firm Segment - usually recently established and innovative. High birth and death rates, but potential for growth.

B. THE LARGE BUSINESS ORGANISATION SEGMENT.

    1. The Multi-division segment. - currently the dominant form, usually multinational. Structure reflecting the classical 'product cycle'
    - hence further classification:

      a. Leaders - Institutional form of growth phase of product cycle. Innovative, high-risk/high-reward.
      b. Intermediate - Institutionalised form of maturity phase of product cycle. Steady but modest profits.
      c. Laggard - Institutionalised form of tail of product cycle. Steady but low profits.
      d. Support - providing general services (including transfer pricing) to multidivisional corporations.

    2. The Global Corporation Segment - those very largest companies involved in the 'new international division of labour'. Structure similar to multidivisional segment, but strategy described as 'opportunist' or 'fiscal instrumentality'.

functions. Product changes will be few, and not science-based. Such firms are likely to be late adopters of relatively simple process innovations, relying on machinery and equipment manufacturers and their salesmen for technical advice.

b. Satisfied. These are the non-ambitious enterprises "often quite purposely kept small" (Taylor and Thrift 1983, 452). The goals of the enterprise are to be understood in behavioural rather than economic terms, and research and innovation are unlikely to be evident.

2. The Intermediate Segment

a. Satellites. Whether subcontractors or franchisees, such firms are dependent on the large firm segment. The expected innovation strategy is that described by Freeman as 'dependent', where the firm innovates only on demand from its customer and to a specification set by its customer. The firm is production-oriented and might thus be expected to be an adopter of well-tried process innovation. In-house R&D is not to be expected, although the firm will be production oriented and will thus require design and production engineering functions.

b. Loyal Opposition. This segment exists outside the competitive sphere of the large corporation and the expected innovation strategy is 'traditional'. Production criteria are again the most important and late adoption of process innovation is to be expected on the basis of little or no in-house R&D.

3. The Leader Firm Segment. These are the innovative small firms in the segment which is likened to the classical 'seed-bed'. According to Taylor and Thrift such firms follow 'offensive' or 'opportunist' innovation strategies. The ability to follow a highly research intensive offensive strategy will depend on the threshold R&D levels in a particular industry (see above), and the overall numbers of firms in this category is likely to be small. 'Opportunist' strategies owe little to formal R&D and more to imaginative entrepreneurship.

B. The Large Business Organisation Segment

1. The Multidivisional Segment

a. Leaders. These are the research intensive organisations operating an 'offensive' or 'defensive' innovation strategy. Such firms will emphasise research rather than development, and those few firms pursuing the 'offensive' strategy will perform fundamental research and thus be linked to world science and technology. Their emphasis will be on product innovation (although some of these may constitute process innovations for other, customer, firms) and they will be operating at the leading edge of the product cycle where technical specifications are not standardised and where production costs and output levels are not critical.

b. Intermediates. These firms are operating at a later stage of the product cycle, content to pursue an 'imitative' innovation strategy and with an increasing emphasis on production criteria. Thus process innovation will be important to the firm, while product innovation will be based on the acquisition of licenses and know-how from 'offensive' and 'defensive' innovators (who may in fact be part of the same multidivisional corporation). In-house R&D will be necessary to implement acquired product and process innovation, but there will be increased emphasis on development and little or no fundamental research will take place.

c. Laggards. These firms are operating at the tail of the product cycle. Products will be standardised and the opportunities for product innovation will be minimal. Production criteria will dominate and there will be an emphasis on cost and labour saving process innovation. Automation and long production runs will characterise this segment. Some such firms will operate 'dependent' innovation strategies where the dependence will often be on other organisations within the same corporation. Other such firms will be 'imitative' and for these the difference from the intermediate segment will be one of degree - the extent of product innovation even more limited and the production orientation even greater. There will be no research, but the adoption of process innovation and of minimal product innovation may require on-site development.

d. Supports. Firms in this sector are essentially non-manufacturing. Innovation will be non-technological.

These suggestions are summarised in Table 3 which shows the sorts of Research, Development and Scientific and Technical functions which might be expected to be associated with each segment of the segmented economy.

Segmentation and Space: Peripheral Economies

Taylor and Thrift (1983, 459) describe an agenda for research which includes: "The investigation of the regional and national distribution of members of particular segments, business organisations and constituent establishments, as an explanation of uneven regional and national development".

Taylor and Thwaites (1981) have given some preliminary attention to the role of technology in the segmented economy, arguing that studies of innovation undertaken at the Centre for Urban and Regional Development Studies provide empirical illustration of the segmented economy in the UK. What is not clear from such an argument is whether the differential levels of innovative performance are seen as a dependent or independent variable, i.e. as the cause or consequence of industrial segmentation.

A different argument is proposed here: since hypotheses can

Table 3. Research Development, Scientific and Technical Functions in the Segmented Economy

SEGMENT	FUNDAMENTAL RESEARCH	RESEARCH	DEVELOPMENT	OTHER SCIENTIFIC AND TECHNICAL FUNCTIONS
SMALLER FIRM SEGMENT:				
Laggards:				
Craftsmen:				Expertise acquired with purchased process innovations
Satisfied:				
Intermediates:				
Satellites:			Some, to incorporate innovation to specification	Design, Production Engineering, etc.
Loyal Opposition:				Expertise acquired with purchased process innovation. Design, production engineering, etc.
Leaders:	Perhaps, but unlikely to be able to afford very much			Technological entrepreneurship and opportunism
THE LARGER BUSINESS ORGANISATION SEGMENT:				
The Multidivisional Segment:				
Leaders:	Yes	Yes	Yes	All scientific and technical services important
Intermediates:		Some	Yes	Design and Production Engineering
Laggards:			Some	Design and Production Engineering. Process innovation bought in.

be generated about the innovative performance and research requirements of firms in different segments of the segmented economy, then the identification of the segmented structure of regional economies (as proposed by Taylor and Thrift) will allow us to conjecture a pattern of regional research performance. Consistent with the demand-side emphasis on innovation proposed above, we will arrive at a functional account of R&D activity in regional economies. The policy implications of this will then need to be explored.

A great deal is already known about the types of firms which occupy peripheral locations. In particular the multidivisional corporate sector in peripheral regions is known to be characterised by 'branch plants', and such plants dominate regional economies. In the Northern region of the UK 73 per cent of all plants were externally owned in 1973 (Smith 1979), while Crum and Gudgin (1978) have shown that such plants have a smaller number of managerial, technical and administrative jobs than do production sites at headquarters in core areas. There is little direct data to show that the nature of production at such plants is characteristic of manufacturing at the end of the product cycle - evidence on skill levels of the workforce, and of capital/labour ratios is not available. Nevertheless, a range of anecdotal and systematic data (Watts 1981) points to the predominance of branch-plant manufacturing by the corporate sector in peripheral regions. Such plants constitute the laggard or intermediate segments of the large business organisation segment in Taylor and Thrift's segmented economy.

Rather less is known about the smaller firm segment in peripheral regions, although the relative absence of leader firms has been noted by (among others) Storey (1983). The Regional Studies Association (1983, 52) reports that "..small firms do not exist as an independent economic sector; in manufacturing the majority are dependent on larger firms within local or regional markets".

While it is far too early to assume that the analysis of segmentation in peripheral economies is complete, it would appear not to violate reality to an unacceptable extent to assume that the manufacturing sector in peripheral economies in the UK is dominated by a small number of segments:

I   Laggard firms - the corporate segment.
II   Intermediate firms - the corporate segment.
III Satellite firms - the intermediate smaller
     firm segment.
IV Laggard firms - the smaller firm segment.

Segmentation, Space and R&D

The implication of the demand-led theory of R&D elaborated in this chapter would lead us to expect a limited range of Research and Development in any economy characterised by the business segments above.

Oakey, Thwaites and Nash (1982) report a pattern of regional variation in product and process innovation in the scientific instruments, electronic components and metal working machine tool industries in the UK. The study distinguishes between group plants and single independent plants, which can be taken to correspond to the multidivisional segment and the smaller firm segment described above. Product innovation in the single plant sector is seen to be far more frequent in the South East than in the Northern region, reflecting a higher incidence of laggard firms in the North. In the group plant sector the distinction is similar but less marked, but it is noted that group plants in the North do not develop these innovations 'in-house' but acquire them via group plants in core regions.

In the case of process innovation little regional differentiation exists. Rates of process innovation are high in all segments in peripheral regions, which again is consistent with the production orientation of intermediate and laggard segments.

Thwaites (1982) reports further on studies of the same industries, and in particular on the incidence of research and development. The incidence of R&D activity at a site is found to be significantly related to the incidence of product innovation. (But the definition of R&D used in this study is not reported, and it is possible that a somewhat tautological relationship is being identified.) Analysis which differentiates between group plants and single independents again shows variations in R&D activity in the expected direction, with lower levels of activity in peripheral regions. Although group plants tended to have some R&D in peripheral regions, the size of the R&D effort was smaller.

While these results are clearly in accord with the theory of segmentation and R&D reported above it should be noted firstly that the results derive from a study of only three MLH headings, chosen because they were known to be highly innovative, and secondly that no distinction is available between different forms of R&D activity. As Table 3 suggests different segments will demand different forms of R&D, varying from fundamental research (perhaps even true 'basic' research) in the case of the leader multidivisional segment through to activities which according to the widely accepted 'Frascati' definitions (OECD 1981) are not to be counted as R & D at all. Thus the introduction of a process innovation, will at a minimum require a technical feasibility study which according to Frascati should not be classified as R&D but is nevertheless a vital technical function in intermediate and laggard segments.

A thorough empirical examination of regional R&D in terms of the segmented economy would demand discrimination between these different forms of R&D and related activities.

Conclusions and Policy Implications

The proposition outlined in this chapter, that a demand-side emphasis on research and innovation, coupled with the existence of a segmented economy, leads to a functional explanation of regional differences in R&D performance, plainly requires further empirical corroboration. There is a danger, too, in the inherent conservatism of any functionalist account. Thus to explain low levels of R&D in the Northern Region of England as being an inevitable consequence of power relations in the segmented economy must not lead to a fatalistic acceptance of the inevitability of these low levels for all time.

It is necessary however to offer policy prescriptions in the light of the functionalist account. If it is true that Northern region firms are satellites, laggards, and intermediates, it is simply insufficient to exhort these firms to do more R&D. Laggards and intermediates in the corporate sector are served through corporate structures with required process innovations; satellites in the smaller firm sector have little or no use for their own R&D capability. What is at issue is the marginal product of R&D and the account given above shows that for most of the segments expected to be found in the peripheral regions this marginal product will be low or even negative. The number of firms with a positive marginal return to R&D may be vanishingly small.

The determination of an innovation policy for peripheral regions therefore takes us away from concern with the individual firm and towards the whole set of industrial purchasing and power relationships. Perhaps the most powerful tool of intervention possessed by central government is that of public sector purchasing (Cabinet Office 1980) where government can have a direct influence on technical standards and thus on the demand for innovation. But many avenues of intervention are available to be explored, and it is the purpose of this chapter not to explore them but to canvass support for the functionalist demand-led view of R&D and innovation which has been presented above.

References

British Business (1983) 'Industry Carried Out R&D Worth £3.8 billion in 1981.' 9.12.83

Business Monitor MO14 (1980) Industrial Research and Development, Expenditure and Employment, 1978, HMSO, London

Buswell, R.J. and E.W. Lewis (1970) 'The Geographical Distribution of Research Activity in the United Kingdom' Regional Studies, 4, 297-306

Buswell, R.J., R.P. Easterbrook and C.S. Morphet (1985) 'Geography, Regions and Research and Development Activity: the case of the United Kingdom' In: A.T. Thwaites and R.P. Oakey (Eds.) The Regional Impact of Technological Change, Frances Pinter, London

Cabinet Office (1980) Research and Development and Public Purchasing, Cabinet Office, London

Carter, C.F. and B.R. Williams (1957) Industry and Technical Progress: Factors Governing the Speed of Application of Science, C.U.P., London

Cmnd 4814 (1971) A Framework for Government Research and Development, HMSO, London

Crum, R. and G. Gudgin (1978) Non-production Activities in UK Manufacturing Industry, EEC, Brussels

Freeman, C. (1982) The Economics of Industrial Innovation (2nd ed), Frances Pinter, London

Froebel, F., J. Heinrich and O. Kreye (1980) The New International Division of Labour, C.U.P., London

Gilpin, R. (1975) Technology, Economic Growth and International Competitiveness. Study prepared for the Subcommittee on Economic Growth of the Congressional Joint Economic Committee, US Government Printing Office, Washington, DC

Langrish, J., M. Gibbons, W.G. Evans and F.R. Jevons (1972) Wealth from Knowledge, Macmillan, London

Mansfield, E. (1971) Research and Innovation in the Modern Corporation, Norton, New York

Metcalfe, J.S. (1970) 'The Diffusion of Innovation in the Lancashire Textile Industry', Manchester School of Economics and Social Studies, 2, 145-62

Mowery, D. and N. Rosenberg (1979) 'The Influence of Market Demand upon Innovation: a critical review of some recent empirical studies' Research Policy, 8, 103-53

Oakey, R.P., A.T. Thwaites and P.A. Nash (1982) 'Technological Change and Regional Development: some evidence on regional variations in product and process innovation' Environment and Planning A, 14, 1073-1086

OECD (1981) The Measurement of Scientific and Technical Activities: Proposed Standard Practice for Surveys of Research and Experimental Development (Frascati Manual), OECD, Paris

Regional Studies Association (1983) Report of an Inquiry into Regional problems in the United Kingdom, Geo Books, Norwich

Rothwell, R. (1982) 'The Role of Technology in Industrial Change: implications for regional policy' Regional Studies, 16, 361-369

Schmookler, J. (1966) Invention and Economic Growth, Harvard University Press

Smith, I.J. (1979)'The Effect of External Takeovers on Manufacturing Employment Change in the Northern Region between 1963 and 1973' Regional Studies, 13, 421-37

Storey, D. (1983) Entrepreneurship and the New Firm, Croom Helm, London

Taylor, M.J. and N. Thrift (1983) 'Business Organisation, Segmentation and Location' Regional Studies, 17, 445-465

Taylor, M.J. and A.T. Thwaites (1981) 'Technological Change and the Segmented Economy' Paper given to the Regional Science Association, Durham, September 1981

Thwaites, A.T. (1982) 'Some Evidence of Regional Variations in the Introduction and Diffusion of Industrial Products and Processes within British Manufacturing Industry' Regional Studies, 16, 371-381

Utterback, J.M. (1974) 'Innovation in Industry and the Diffusion of Technology' Science, 183, 620-6

Watts, H.D. (1981) The Branch Plant Economy: A study in external control, Longman, London

Williams, B.R. (1967) Technology, Investment and Growth, Chapman and Hall, London

CHAPTER FOUR

STRATEGY FORMULATION, ORGANISATIONAL LEARNING, AND LOCATION

Dr. J.W. Harrington, Jr., Assistant Professor Department of
Geography, University Buffalo, USA [1]

Introduction

The presence, growth, or decline of economic activity in a
particular industry within a given region has an effect on the
region's economy that differs according to the nature of the
activity and the characteristics of the organisations engaging in
the activity. This chapter focuses on differences among
enterprises operating in the same industry that may affect the
locational needs and impacts of their varied activities. First,
hypotheses are developed to relate geographical rationalisation
of activities to differences among enterprises in the same
industry. Then, published empirical work and unpublished case
studies are used to illustrate the relationships posited, within
a variety of industrial and geographical contexts. Case materials
were compiled from the author's investigations of the US-based
semiconductor-device industry during the 1960-1980 period.

Activity Rationalisation

Industrial geographers have expressed concern for the
differential impact of varied activities within the same
enterprise and/or industry (Westaway 1974b, Clark 1981). It is
now understood that, if not how, internally differentiated
enterprises may geographically rationalise their control,
operational, and marketing activities at regional, national, or
international scales. This rationalisation is motivated by
differences in the locational and cost constraints upon the
varied activities, by geographical procurement and marketing
strategies, and by changes in technical and market
characteristics of the industry. Experienced industry actors
learn to organise each stage of production in its most efficient
manner. Given sufficiently high value-to-weight ratios of
material movements, efficiency may call for these stages to
separate geographically, allowing the locational factors
important to each stage to affect that stage's location more

63

directly. Rationalisation is constrained by the organisational and communications capabilities of the particular enterprise. However, the extent and nature of this rationalisation is not a function of enterprise size and managerial sophistication only. Enterprises' geographical structures differ because of conscious decisions and historical attributes of structure, strategy, and ability to learn from their experiences and environments.

Intraindustry Differentiation: Strategy Formulation and Organisational Learning

Organisation studies generally consider two types of differences across enterprises: structure and strategy. Organisational structure, the system of oversight, directive, and reporting relationships among and within functional units, is influenced by and influences local economic and social characteristics. Structure may "reflect spatial variations in environmental conditions" (Marshall 1982, 1673), including supply of materials, labour, or technical skills, local market structure, and market uncertainty. Structural characteristics of dominant enterprises in turn affects the occupational structure and quality of work life in local areas. Within a multisite enterprise, structure generally has spatial manifestations in the differentiation of personnel and activities across sites and the nature of intersite linkages. These linkages affect the local economic and technical impact of the enterprise (Westaway 1974a and b, McDermott and Taylor 1982, Ch. 3). Of course, the correspondences between internal organisation and locational organisation are not always direct. While enterprise case studies benefit from attention to this correspondence, it is difficult to generalise organisational structure in space without more information about the enterprise.

Since Chandler's (1962) pioneering work, theorists and planners have viewed structure as a manifestation of organisational goals, activities, and environment (suppliers, competitors, and markets), all of which reflect the organisation's strategy. Thus, strategy becomes the variable to be studied to understand differences among enterprises. Strategy is the means by which an enterprise or business unit matches its resources to the requirements of its environment (Hofer and Schendel 1978). This match is accomplished by modifying the resources available to the enterprise, modifying the environment, or changing the enterprise's operations to a different (part of its) environment. Thus, strategy can be analysed along as many dimensions as there are possible enterprise actions: modifications of products, production, markets, or organisation. Certain aspects of strategy largely determine elements of organisational structure, and structure can be used as a proxy for the strategy. For example, the geographic and product markets in which an enterprise is involved may be reflected in a geographical product-line structure. In some cases, though, the connection is not so clear-cut, and strategy cannot be inferred from structure.

Differential bundles of fixed assets and of target markets allow a range of 'optimal' strategies at the corporate (the range and relationships of sectors in which the enterprise is involved) and business (competitive role within an industry) levels. Within a single industry, enterprises may pursue different strategies because of: heterogeneity of the fixed assets (skills, experience, organisation, physical capital) associated with the enterprises; differences in the locations of those assets; the imperfection of markets for exchange of these assets; and the heterogeneity of buyer preferences in some markets (Newman 1978, Caves 1980, 64-65). The resources and markets sought by and allocated within the enterprise are spatially variable, and the enterprise's internal allocation must be spatially manifested. Therefore strategy, along dimensions appropriate for the industry to be investigated, may be a key to the enterprise's geographical distribution of activities and resultant local impacts.

Another important enterprise-specific characteristic is its plans and capacity for improving organisational performance through experience. Organisational learning is the regular, internalised generation of knowledge specific to the needs of an enterprise, knowledge subsequently appropriated by the enterprise (or components thereof). Learning has occurred when an action has been conditioned by evaluation of the results of a previous action as well as by assessment of the relevant environment for the action. Enterprises may take measures to increase the likelihood of successful learning (Argyris and Schon 1978, Day and Tinney 1968). The types of measures depend upon the object of the experience-based learning. Employee-embodied learning to reduce direct-labour requirements (the 'learning curve') requires minute division of labour, employee tenure, strict supervision, and progressive mechanisation (Abernathy and Wayne 1974). Strategic learning -- interpreting, reacting to, and modifying the business unit's environment -- requires monitoring of environmental changes and of business performance, awareness of the internal roadblocks to more than minor changes, and the ability to change more than routine operating behaviour (Argyris and Schon 1978).

The type of information to be gathered influences the appropriate contractual and organisational mode. Practical knowledge may be supplied by machinery, written instruction, or consultant service. Gaining hand-on ability requires demonstration and repetition. Finally, the target population within an enterprise influences the channels of information transfer used. For example, Havelock (1969) suggested that reading and lectures are more appropriate for dissemination of factual information to highly educated groups. These influences suggest that the occupational distribution, organisational structure, and even the internal geography of an establishment should reflect the types of information-gaining needs perceived by the establishment and its enterprise.

The ability to learn is not uniform across enterprises, and thus represents a distinguishing feature for the abilities of enterprises to rationalise production, markets, and their locations. It also represents a source of heterogeneity of the locational needs and local impacts of enterprise activities. Activities and enterprises organised around learning attempt to hire and retain more skilled workers, and tend to cluster spatially to capture external economies of skill and experience.

Hypothesised Relationships

With these relationships as a conceptual basis, several hypotheses can be developed regarding strategic differentiation within industries, strategic formulation within enterprises, and locational distribution of the activities within an industry. Each hypothesis is illustrated and further developed in turn.

1. Competition among business units within an industry is tempered by strategic differentiation along the dimensions of scale of operations, breadth of product line, breadth of internalised functions, cost and pricing posture, and innovativeness.

In a study of new business ventures by established companies, Biggadike (1979) found that ventures entering with a high share made much higher profits sooner than low-share entrants. Ventures that saw rapid growth in share after entry made the lowest profits of all. The relevant measure of share was relative share, "the venture share divided by the combined share of the three largest competitors" (p.542). Other size measures that were strongly correlated with profit were production scale (initial production capacity divided by market size) and two measures of market scale (number of products offered as a percentage of the product range of the largest competitors and the number of customers as a percentage of customers of major competitors). The salient break between low- and high-profit earners was 20 per cent production scale, 100 per cent product-market scale, and 33 per cent customer scale. Day (1975) also found relative share to be important. He found a clear link between share and profitability in business units of large US companies, if a unit's share is at least fifty per cent larger than its nearest competitor.

Enterprises' attitudes toward and organisation for stringent cost reduction also segment many industries. Abernathy and Wayne (1974) counterposed adherence to experience-based production-cost cutting versus production flexibility and innovativeness as two very different strategies. The learning curve must be managed by product standardisation, financial capital tied in vertical integration and specialised equipment, reduction of labour input, increased scale, and increased labour specalisation. In many industries, the potential for rapid cost reductions is slim. In Biggadike's study of new ventures, all ventures faced production, marketing, and development costs that remained high during the first four years. Profitable ventures (quite the minority) benefitted from rapid increases in revenues rather than any sharp reduction in costs.

In the early 1960s Texas Instruments led the semiconductor industry in sales, and its technology was considered second only to Western Electric/Bell Laboratories. During the 1970s, TI's avowed strategy was to continue as a cutting-edge product innovator, while the company gained the reputation of a manufacturing giant, focused on production technology and scale. By using the learning curve concept, TI realised the importance of being first to introduce a product to the market and maintaining a leadership position in high-volume markets. For example, in 1968 TI announced its 64K memory several months ahead of its competitors. During the mid-1970s, TI successfully competed with large Japanese producers as prices for simple watches and pocket calculators plummeted, driving several large US producers from these low-end consumer markets. By 1978 programs were underway to automate the assembly of calculators, large scale integrated circuits, and a broad range of other products (Busness Week 1960, 1978).

2. Strategic differences among business units are reflected in the units' geographical structures and in the places in which the enterprises locate specific activities.

The potential importance of corporate and business strategy for the geographical structure of enterprises is a key to intraindustry differentiation within industrial geography. Harrington (1985a) identified three dimensions of strategy with implications for the locational needs of control, development, and operations activities of semiconductor-device manufacturers. The first dimension, the use of vertical integration as a corporate strategy, influenced the extent and location of geographical rationalisation of activities. Between 1965 and 1980, US companies founded specifically to develop and produce semiconductor devices established many more foreign device-assembly facilities than did pre-existing equipment manufacturers that innovated into or integrated backward into semiconductor manufacturing. Conceivably, the internal markets of integrated producers shielded them from some of the cost-cutting and market-protection pressure that motivated the independents' moves. Among pre-existing companies, the older electronics companies developed more domestic branches and greater geographical separation of control, research, design, and manufacturing facilities than did the more widely diversified companies.

Since the early 1960s, TI has been unusually vertically integrated. Its production engineering group designed and produced most of the company's equipment. The company's vertical movement was also into raw materials, instruments, and other electronic devices. Its silicon plant was built to supply silicon for other manufacturers as well. The strategy of vertical integration was extended when, in 1972, TI successfully established itself in the consumer electronics business with the hand-held calculator. To maintain high production volumes on growth products, without pausing for new plants to gain high productivity, TI built new facilities before capacities were reached at existing fabrication plants. The company's US operations were concentrated in central Texas, and it established fewer Southeast Asian assembly operations than did most of its large US independent competitors. TI viewed its focus on automation as the only way the company could build its price-sensitive products in the US (Business Week 1960, 1978, Texas Instruments 1961, Harrington 1985a.)

Among the 21 semiconductor manufacturers studied, the dimensions of business strategy that distinguished locational patterns were market share (dominant versus small-share, measured across US and world markets for semiconductors of all types) and breadth of product line (specialist in one or two types of semiconductor devices versus generalist across a full range of devices). The group of large-share generalists most actively developed assembly-only facilities in low-wage countries for export to US and other major markets. TI, however, one of the largest and broadest producers during the period, maintained a significant US presence across all activities, partly due to vertical integration and partly via process automation.

Most specialist companies concentrated in higher-performance and higher-margin products, leaving generalists with greater concern for costs at the lower end of the market. National Semiconductor emerged from a change of financing and management in 1966 with a goal of manufacturing efficiency and high volume production of a wide range of standard products. National's rapid growth and cost cutting emphasis were reflected in the number, locations, and nature of its facilities. In 1970 the company had five facilities in four countries and had assembly subcontractors in three Southeast Asian countries and in Mexico. By 1975 thirteen wholly-owned operations were in ten countries. As of 1980, the company owned nineteen facilities in twelve countries. In contrast to Intel, many of National's facilities were assembly plants in Southeast Asia. During the same period, marketing operations were established in two European countries, with an additional fabrication and testing facility in Scotland.

With respect to breadth of functions within the organisation, Harrington noted locational differences between semiconductor companies focused on a few activities and those that provided a broader range of activities internally. Greater internal differentiation, especially when combined with broader product lines, encouraged greater geographical differentiation (by location and by type of place) of varied activities within companies. Scott (1983a,b) posited that the degree of vertical integration within an enterprise, or dominant within an industry as a whole, is directly related to the degree of spatial deconcentration of the industry within a metropolitan area. This vertical (dis)integration may occur at such fine operational distinctions that customary boundaries between industries are not crossed.

3. Local impact of enterprises' operations is influenced by the business strategy of the enterprises.

The local impact of an enterprise's operations should reflect the corporate and business strategies of the enterprise. The occupational mix of particular operations is influenced by the organisational structure and the degree of geographical differentiation of the enterprise. These are in turn influenced by business strategies of share and scope, as presented above. The occupational mix of major local employers helps determine the range of local job opportunities, social characteristics, and salary/wage mix. In rapidly changing industries, the activity and occupational mix at a particular establishment affects the probability of local entrepreneurial spin-offs or purchases of the establishment, each a strategic reaction to industry change.

Intel built its reputation on technical breakthroughs, from its start in 1968. In the early-to-mid 1970s, Intel pioneered first the computer memory chip and then the microprocessor. Innovation and concentration in semiconductor memory and microprocessors gave Intel short technological leads. In the mid-1970s, the company began to stress manufacturing and marketing skills to achieve sales increases in large, growing, and hotly contested markets. Intel used major capital expenditures in 1978-1980 to deconcentrate its design, fabrication, and assembly operations outside of California. To facilitate information transfer within innovative product divisions, Intel tended to relocate divisions as coherent units, including design and planning functions (Business Week 1980b). New locations thus required a range of occupational skills and ancillary services. By the late 1970s, Intel's major operations existed in Santa Clara County, and near Portland, Oregon, and Phoenix, Arizona. The company's decision was likely influenced by the abundance of technical and assembly talent in Phoenix. Other semiconductor companies already in the Phoenix area (Motorola, General Instrument, EMM Semi Inc., General Semiconductor, Northern Telecom, and Medtronics) had created a substantial labour pool (Waller 1979). In 1980, Intel opened a design centre in Israel and was planning another in Japan to tap new sources of engineering talent. The company publicly expressed the desire to find new communication techniques that would tie the organisation together (Business Week 1980a).

Local stability of corporate control and support employment

depends in part upon corporate earnings stability. In the medium run (across the business cycle) this stability reflects the corporate mix of markets and industries and corporate strategy for growth (see Krumme and Hayter 1975). Operations employment stability depends upon the nature of remote operations -- duplicative versus integrated versus local-market-oriented -- and the use of subcontracting by the enterprise. Subcontracting is widely used in some industries to absorb fluctuations in demand facing primary producers. In other industries subcontracting is used more consistently to fulfill some stage(s) of production. Scott (1983a, 242-6) posited that local subcontracting linkages will be greater for small and fluctuating contracts, and found (1983b, 359-60) that smaller printed circuit board manufacturers were much more likely to subcontract locally. Taylor and Thrift (1982, 1623-4) found local subcontracting to be widely used only by smaller, technologically backward enterprises. The local impact of subcontracting is a higher but more volatile employment multiplier for the primary enterprise's activity.

With respect to local linkages, Averitt (1968) suggested that smaller enterprises have local supply sources and local markets (the latter, especially if advertising is important). In a survey of iron boundaries in the West Midlands, Taylor and Thrift (1982) found the local sales linkages of smaller, independent enterprises to be much greater than those of large enterprises or enterprises closely associated with large enterprises. Local purchasing linkages were strongest among associated enterprises and the technological- and sales-lagging business units of large enterprises.

4. Enterprises' organisational structures and geographical structures influence their abilities to learn about their environments and to modify strategy accordingly.

Internal information flows that encourage strategy formulation and change have been found to increase with geographical propinquity of an enterprise's activities, multiplicity of work relationships within the enterprise, identification with and esteem of participants within the enterprise (Davis 1953), perceived value of the information, and a sense of crisis within the organisation. A perceived need for stability, due to large fixed-capital investment, vertical integration, or managerial style, inhibits information flow and strategic change (Havelock 1969, Abernathy and Wayne 1974). While the geography of the enterprise does not seem to be a major determinant of information flow, a number of studies have given propinquity some importance (Davis 1953, Barnlund and Harland 1963). One might expect that business units (and parts of business units) whose strategies entail the most intensive information-generating and information-gathering require the most geographical clustering of operations. At the local level this clustering brings a range of activities and occupations, and also requires a minimal information and labour-force availability. For industries and enterprises in which subcontracting is common, local availability of subcontractors should be important for information-intensive operations.

Concurrent with the ownership and managerial changes in National Semiconductor in 1966 were changes in product line and headquarters location. The company turned form discrete transistors to integrated circuits, making a five-year-late transition in new product technology. The company's headquarters and additional facilities were relocated from Connecticut to Santa Clara County, California. In 1980 National announced its decision to locate a water fabrication facility for bipolar integrated circuits in Arlington, Texas (between Dallas and Fort Worth). The reason given for the company's selection of the site included the area's stable community, which was attractive to technical professionals, its suburban atmosphere, and its proximity to outstanding educational facilities (Ferguson 1980). Clearly important, as well, were the labour resources and supplier networks created by TI's bipolar facility and the headquarters and operations of Mostek, a TI spin off, all within 15 miles of Arlington.

Conclusions

The development of the hypotheses in this chapter and the cases provided illustrate the complex and contingent nature of the relationships postulated. The cases make clear the extent to which three large companies maintained their success during the study period by pursuing, obtaining, and investing in various sources of monopolistic advantage, even while producing semiconductor devices that are marketed nearly as commodities. The search for enterprise-specific advantages created strategic groups within the industry, each group led by the strategy's most successful proponent. The organisational structures, R&D policies, and process technologies of these companies were influenced by their dominant strategies of large-share generalist (TI), large-share specialist (Intel), and large-share, low-cost hybrid (National). (These descriptions are supported by 1980 global market share estimates by product line (Dataquest 1984)). In addition, the mix of locations selected for various corporate and business activities reflected the cost, flexibility, and information needs of the companies' dominant strategies. Given the technologies and geographical cost disparities available during the study period, the most severe cost cutting was accomplished with the assistance of offshore assembly operations. Product flexibility and information transfer were enhanced by comprehensive facilities in places large enough to support their size, while production-level flexibility was achieved via layoffs. Rapid development of a comprehensive facility was assisted by location in a local labour market containing existing semiconductor operations and in a region containing electronics supplies and services.

To the extent that these findings are generalisable, they suggest a level of analysis to be added to the spatial division of labour already considered by industrial geographers. Industrial locational analysis should be executed with an awareness of industrial characteristics (including competitive structures and bases of competition), changes in those characteristics, differences in the locational influences relevant to particular activities (stages within the production sequence as well as functions within enterprises) within the industry, and differences in the locational influences relevant to particular enterprises within the industry. Enterprises may be grouped according to characteristics (termed strategic characteristics in this chapter) that suggest the differences in locational influences. Thus, the difficulties of considering each enterprise to be sui generis are supplanted by the difficulties of identifying the dimensions of strategy most relevant to the locational analysis in a given industry. The case studies used in this chapter suggest that vertical integration, product and process technological focus, market share, and market position may generally be significant dimensions of enterprise strategy.

Measurement of actual local impact requires further research. Currently, the results of the differences outlined above can only be implied. Comprehensive facilities should require a greater range of skill and pay levels within the workforce. Automation of manufacturing processes tends to increase division of labour, skill specialisation, and organisational complexity (Blau et al. 1976), increasing the heterogeneity of the workforce. In this industry output flexibility tended to be achieved by maintaining multiple operations whose output and employment levels fluctuated widely (as opposed to subcontracting, for example). Now that many semiconductor operations of various kinds have operated outside of the industry's core regions for some time, inspection of local employment, wage, and linkage impacts would be a valuable exercise. Cross-industry studies of this type would eventually allow analysis of industrial impact more useful to a specific region than analysis based on implications drawn from industry-wide characteristics.

This chapter has attempted to augment a conceptual background for such investigations, by focusing attention on types of differences among business units and among their activities within an industry. These differences in strategy and learning capability affect the manifestation of competition in the industry and operate within the changing constraints of the industry's evolving technology and markets. As such, they are an important link to the local, human outcomes of global competitive struggles within and among industries.

Note

1. The author acknowledges the invaluable assistance of Christopher Montante in drawing together the case studies.

References

Abernathy, W.J. and K. Wayne (1974) 'Limits of the learning curve' Harvard Business Review (Sept. Oct.), 109-19

Argyris, C. and D.A. Schon (1978) Organizational Learning: A Theory of Action Perspective. Addison-Wesley, Reading, MA

Averitt, R.T. (1968) The Dual Economy: the Dynamics of American Industry Structure. Norton, New York

Barnlund, D.C. and C. Harland (1963) 'Propinquity and prestige as determinants of communication networks' Sociometry, 26, 467-79

Biggadike, R. (1979) 'The risky business of diversification' Harvard Business Review (May). Reprinted in M.L. Tushman and W.L. Moore (Eds) Readings in the Management of Innovation. Pitman, Boston

Blau, P.M., C.M. Falbe, W. McKinley and P.K. Tracy (1976) 'Technology and organization in manufacturing' Administration Science Quarterly, 21, 20-30

Business Week (1960) 'Semiconductors' 26 March, 106-10

Business Week (1978) 'Texas Instruments shows US business how to survive in the 1980s' 18 September, 66-90

Business Week (1980a) 'Intel: the microprocessor champ gambles on another leap forward' 14 April, 93-103

Business Week (1980b) 'More elbowroom for the electronics industry' 10 March, 94-100

Caves, R.E. (1980) 'Industrial organization, corporate strategy, and structure' Journal of Economic Literature, 18, 64-92

Chandler, A.D. (1962) Strategy and Structure. MIT Press, Cambridge

Clark, G.L. (1981) 'The employment relation and spatial division of labour: a hypothesis' Annals of the Association of American Geographers, 71, 412-424

Dataquest, Inc. (1984) Preliminary Market Share Estimates (January 27)

Davis, K. (1984) 'Management communication and the grapevine' Harvard Business-Review, 43-49

Day, R.H. and E.H. Tinney (1968) 'How to cooperate in business without really trying: a learning model of decentralized decision making' Journal of Political Economics, 76, 583-600

Ferguson, J. (1980) 'National picks Texas for IC site: volcano kills Washington plans' Electronic News (25 August), 32

Harrington, J.W. (1985a) 'Corporate strategy, business strategy, and activity location' Geoforum, 16, 44, 349-357

Harrington, J.W. (1985b) 'Intraindustry structural change and location change: US semiconductor manufacturing, 1985-80' Regional Studies, 19, (4), 343-353

Havelock, R.G. (1969) Planning for Innovation through Dissemination and Utilization of Knowledge. Institute for Social Research, Ann Arbor, MI

Hofer, C.W. and D. Schendel (1978) Strategy Formulation: Analytical Concepts, West, St. Paul MN

Krumme, G. and R. Hayter (1975) 'Implications of corporate strategies and product cycle adjustments for regional employment changes' In: L. Collins and D.F. Walker (Eds) Locational Dynamics of Manufacturing Activity, Wiley, London

Marshall, J.N. (1982) 'Organisational theory and industrial location' Environment and Planning, A 14, 1667-83

McDermott, P. and M. Taylor (1982) Industrial Organisation and Location. Cambridge University Press, Cambridge

Newman, H.H. (1978) 'Strategic groups and the structure-performance relationship' Review of Economics and Statistics, 60, 417-27

Scott, A.J. (1983a) 'Industrial organization and the logic of intrametropolitan location, 1: theoretical considerations' Economic Geography, 59, 233-50

Scott, A.J. (1983b) 'Industrial organization and the logic of intrametropolitan location, 2: a case study of the printed circuits industry in the greater Los Angeles region' Economic Geography, 59, 343-67

Taylor, M.J. and N. Thrift (1982) 'Industrial linkage and the segmented economy: an empirical reinterpretation' Environment and Planning, A 14, 1615-32

Texas Instruments (1961) Annual Report

Waller, L. (1979) 'Intel sets up plants in Phoenix' Electronics (12 April): 50, 52

Westaway, J. (1974a) 'Contact potential and the occupational structure of the British urban system 1961-1966: an empirical study' Regional Studies, 8, 57-73

Westaway, J. (1974b) 'The spatial hierarchy of business organisations and its implications for the British system' Regional Studies, 8, 145-155

CHAPTER FIVE

ENTERPRISE AND THE PRODUCT-CYCLE MODEL: CONCEPTUAL AMBIGUITIES

Dr. M. Taylor, Senior Research Fellow, Department of Human
Geography, Australian National University, Canberra, Australia

Economic geography has always been long on fact and short on
theory, description has been its forté (Taylor 1984, 1985). It
has and does borrow from other disciplines, especially economics,
the conceptualisations and prescriptions of processes that it
needs to transform descriptions of spatial patterns into more
meaningful explanations. The borrowing began with Thünian and
Weberian location theories and has expanded to include such
diverse areas as dependency theory and export-staple theory.
        The product-cycle model is one of the latest additions to
the list of borrowed explanatory frameworks. It is now being used
more and more extensively in economic geography as an explanation
of changing patterns of economic activity. It is used to explain
the shift of manufacturing from developed to less developed
countries, together with the creation of the New International
Division of Labour (NIDL), and the shift of manufacturing at the
regional scale such as from the Snow Belt or Rust Belt to the Sun
Belt in the USA. It is invoked in studies of individual
industries (for example Auty's (1984) study of the global
petrochemical industry) of analyses of capital markets (Gertler
1985). It is the cornerstone of studies of new firms and high
technology firms (Oakey 1984) and the spatial aspects of high
technology industry in general (Hall 1985). In short, the
product-cycle model is rapidly being elevated by geographers into
a universal explanation of changing industrial location and
spatial uneven development. The rationale for such a move is only
too clear. The product-cycle model refers explicitly to
multinational corporations and multinational capital which
supercedes simpler notions of the firm used in earlier theories
of locations, and the model also emphasises the role of
technological change in the creation of spatial uneven
development.
        The model that was originally proposed by Vernon (1966) and
Hirsch (1965, 1967) was never so ambitious. Indeed, Vernon (1979)
has contended that the usefulness of the model has declined with
time, owing to the greatly expanded multinational network that has
developed in the 1970s and 1980s (including the now global reach

of a large number of enterprises), and the greatly expanded number of countries (including Third World countries) in which multinational corporations network that has developed in the 1970s and 1980s (including the now global reach of a large number of enterprises) and the greatly expanded number of countries (including Third World countries) in which multinational corporations now originate. For Vernon, no more than strong traces of the product-cycle sequence now remain, but for others (for example Auty (1984) and Tichy (1985)), no such dilution of the model's potency has occurred. Thus, Auty (1984, 336) has remarked that: "...if the analytical focus shifts from the advanced economies to the international economy, and a market dynamic complements scale economies, research and development, and labour costs as determinants of global locational shifts, then the product life-cycle model retains its utility." By inference, moreover, the utility of the model remains for a wide range of industries including capital intensive and scale sensitive industries such as petrochemicals.

But then, geographers are not good borrowers of explanatory frameworks. They can be accused of simplifying and distorting the concepts they borrow, not least the product-cycle model. They have often ignored the limitations of the models they acquire and have tended to employ them too extensively, sometimes universalising them. They rarely test these models (like the product-cycle model), but rather invoke them as explanations of patterns that have simply been described. Explanatory models used in geography, therefore, gain currency by assertion rather than by verification and testing. The models also become more and more elaborate until they are naively universalised. The use and abuse of Weberian location theory is a classic example of this process and the product-cycle model is in danger of being accorded the same treatment.

The purpose of this chapter is to counsel caution in the use of the product-cycle model in geography. When the model was proposed by Vernon in the mid 1960s, he attached important caveats to it. These, unfortunately, appear now to have been all but ignored. The product-cycle model also espouses a very particular view of the way in which national, regional and international economies operate. It also contains important assumptions on the functioning of industrial enterprises and on the nature, operation and consequences of technological change in economic development. This combination of viewpoint and assumptions is not always appreciated by geographers when they use the product-cycle model even though they are vital not only to its functioning but also to its applicability. The realism of the assumptions the model makes are, in fact, the key to its usefulness. Furthermore, the product-cycle model has been used in a very particular way in geography. Geographers have emphasised the spatial dimension of the product-cycle and it can be suggested that a distortion has been introduced into the model which closely resembles the distortion introduced into Perroux's

(1950) growth pole theory when it was translated into a spatial regional planning model by Boudeville (1966). What works in flexible economic space does not always translate into less forgiving geographical space.

These problems associated with the product-cycle model are canvassed in the separate sections of this contribution. In the first section, the nature of the product-cycle model is examined together with the elaborations that more recently have been proposed. In the second section, the limitations and assumptions of the model are explored, and in the final section the problems associated with the use of the model in geography are addressed. The arguments presented draw heavily on and extend an earlier critique of the product-cycle model (Taylor 1986).

The Product-Cycle Model

The product-cycle model is an elegantly simple construct which in large measure accounts for its popularity. It was developed in the first instance to explain US trading behaviour and draws together explanatory devices developed in economics since the 1930s: endogenous innovation, the gradual mobilisation of new product technology, the role of information and supply and demand factors (Tichy 1985). The engine of the model is invention, innovation and technological change. In the original model, the USA was seen as the prime location for the inception of the cycle since there high income levels generated demand for a wide variety of sophisticated goods and scarcity of labour encouraged the search for labour saving devices. The model is summarised in Fig. 1 and Table 1.

In the early stages of the manufacture of a new product, the technology is unstable and experimental, demand is uncertain, and technical and scientific ability are vital for survival. The innovating firm needs good and flexible communications with suppliers, customers and even competitors. Price elasticity is low and, within limits, the producer of the new product can set prices without undue concern for competitors. The conditions conducive to this initiating phase of the product-cycle are judged to exist in the metropolitan centres of advanced, industrialised countries, especially the USA. This is, in fact, a natural conclusion for Vernon to have reached given his earlier work on agglomeration and external economies in the New York Metropolitan area (Vernon 1960).

As demand for the product expands both at home and abroad, the growth phase begins, heralding the first move towards product standardisation. Mass production techniques are introduced and the importance of production costs and economies of scale begins to grow. Foreign markets had been served through exports in the earlier phase (Fig. 1), but now the advantages to be gained from such an arrangement begin to diminish. The need for flexible input sourcing declines, process economies become available and,

77

Product-cycle model

Figure 1. The product cycle model

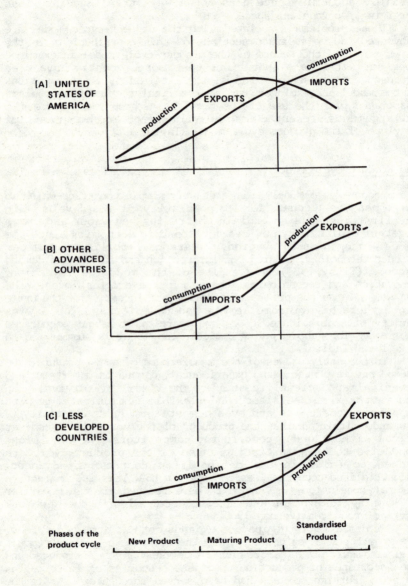

Source: Vernon (1966)

Product-cycle model

Tabel 1. Characteristics of the product-cycle model

CHARACTERISTICS	PHASES OF PRODUCT-CYCLE		
	NEW PRODUCT	MATURING PRODUCT	STANDARDISED PRODUCT
Technology	– Short runs – Rapidly changing techniques – Dependence on external economies	– Mass production methods gradually introduced – Variations in techniques still frequent	– Long runs and stable technology – Few innovations of importance
Capital Intensity	– Low	– High, due to high obsolescence rate	– High, due to large quantity of specialised equipment
Industry Structure	– Entry is know-how determined – Numerous firms providing	– Growing number of firms – Many casualties and mergers – Growing vertical integration	– Financial resources critical for entry – Number of firms declining
Critical Human Inputs	– Scientific and engineering	– Management	– Unskilled and semi-skilled labour
Demand Structure	– Sellers' market – Performance and price of substitute determine buyers' expectations	– Individual producers face growing price elasticity – Intra-industry competition reduces prices – Product information spreading	– Buyers' market – Information easily available

After Hirsch 1965, Walker 1979

though there is a new concern for production costs, price
competition may not have begun. Locational shifts begin in this
phase. Within the advanced industrial countries production may be
decentralised from the initiating agglomerations and overseas
production may begin to develop which may, in turn, expand to
serve markets in third countries. In this phase, there is also a
growing number of competing companies which brings additional
pressures to set up lower cost production facilities as profits
are squeezed. Now there are competitors both at home and abroad
who not only imitate the new product but also improve it (Walker
1979).

Beyond the stage of growth and the maturing of a product
within the cycle, is the stage of standardised production. Mass
production techniques are the norm at this stage, there is strong
price competition and production costs are the dominant influence
on comparative advantage. There is little further innovation
beyond superficial modification of the basic product to achieve
product differentiation with the market. Thus, production becomes
capital-intensive rather than labour-intensive and is moved to
less developed countries for three possible reasons:
1) the stability of the market allows countries with only poorly
   developed marketing skills to specialise in such products;
2) some enterprises will try to maximise production cost savings
   through the use of cheap labour; and
3) the governments of capital-scarce less developed countries
   provide attractive investment grants.
With the internationalised production of standardised products,
the country where the product was first created now begins to
reimport rather than export in this stage of the cycle (Fig. 1).

This is the basic product-cycle model as it was proposed by
Vernon (1966) and Hirsch (1967). However, it has been considered
by some to be incomplete and has been extended to incorporate a
final stage of decline or eclipse (Auty 1984). In this final
stage of the product-cycle, the markets are either slow-growing
or contracting and competition decreases. There is still an
emphasis on the use of unskilled and semi-skilled labour in
connection with mass production technology. Industries begin to
concentrate and vertical integration develops as companies
internalise formerly externally supplied services. Spatial
strategies may become introverted and based on import
substitution and protection rather than on exports and many
industries will become state-owned, state-subsidised or at least
sheltered by the state. This erects barriers to exit in the
economy which at the same time function not only as barriers to
entry but also as barriers to any form of structural economic
change.

The simplicity of the product-cycle model as it has been
outlined above is both its strenghth and its weakness. The basis
of the model is technological change and the invention and
innovation of new products within an international context. With
technological change as the prime-mover in investment decision-

making, the foundation of the model is technological determinism. Within the basic model, supply, demand, enterprise and other aspects of the economic system are subordinated to the technical demands of producing goods that are becoming increasingly obsolescent. Only in the eclipse phase added by later researchers does organisation become dominant over technology. But, this is not a logical progression from Vernon's original product-cycle model. It is more an inappropriate addendum which conflates growth models of firms with just one phase of the extended product-cycle without reconciling those same growth models with all the other phases of the product-cycle. As such, the addition of an eclipse phase to the original product-cycle model can be looked on as an example of the problems of borrowing and extending explanatory models that has typified so much work in economic geography.

Within the basic model, the market is always assumed to be able to absorb all that is produced, even in the face of increasing competition. The only location specific advantages for the supply of inputs are seen as the external economies of scale that are said to be available in agglomerations. An implicit assumption of the model is that inventions are always created in their final form and never evolve in association with their markets as they progress through the cycle. Also, all products are seen to evolve towards eventual and inevitable standardisation. As a result, there is an assumed drive to achieve economies of large scale production which leads invariably to the labour intensive mass production of these standardised goods. Technology, therefore, is seen as moving in only one direction.

The original model is also primarily sector-based and treats individual products quite separately. Substitution and cross-subsidisation between products play no part in the model. More recent proponents of the model, especially those advocating the addition of an eclipse phase, have emphasised the enterprise context within which the model must operate. In particular, they have stressed the context of the multinational corporation. However, in the original formulation of the model, this context was never made clear.

ʾAdding these limitations together, the product-cycle model is perhaps most appropriately described as disembodied, unilinear, technological determinism: 'disembodied' owing to the emphasis placed on the inevitability of standardisation and the search for scale economies through mass production; 'technologically deterministic' because of the primacy accorded to technological change. The next section of this chapter will elaborate the specific criticisms that can be levelled against the model.

The Conceptual Ambiguities of the Product-Cycle Model

Eight inter-related criticisms can be levelled at the product-cycle model which reflect the ambiguity that surrounds a number of its key concepts. These criticisms concern the nature of business enterprises, invention, innovation and the sorts of products that progress through the cycle. They embrace problems of scale of production, location and the availability of external economies of scale. They also involve assumptions about market stability and product standardisation, the relationship between the product-cycle and other cycles that have been recognised within economies at a range of scales, and the preoccupation of the model with only product technology. All these problems combine to undermine the usefulness of the model and the extent of its applicability.

The ambiguous enterprise context

Within Vernon's version of the product-cycle model, ownership and the structure of enterprises and business organisations have virtually no part to play. The control of technology and technology transfer is based on the control of information and proprietary technical knowledge. Control of this knowledge is thought to decline as the cycle progresses and the maturing product can, therefore, be manufactured in one of two ways:

either (a) by foreign producers who have gained the necessary know-how, often backed by their governments in a bid to achieve import substitution;

or (b) by the enterprise that invented the product going multinational to protect its proprietary knowledge in accordance with the postulates of internalisation theory (Coase 1937, Caves 1982, Williamson 1975,1981).

Thus, when the country where a product originates begins to re-import, is this inter- or intra-company trade for which the motivations may be entirely different? Equally, who is doing the exporting from new overseas locations, indigenous enterprises developing new trade and foreign exchange earnings for their own countries, or foreign multinationals extending into regional markets from off-shore production bases?

This mix of enterprise types is fundamental to the functioning of the product-cycle model, but it is never addressed, let alone reconciled, within the model. It was certainly recognised by Parry (1975) in his study of the product-cycle in relation to the UK pharmaceuticals industry, but it was not seen to be problematic. The proponents of an eclipse phase to end the model's cycle certainly recognise this enterprise context (Auty 1984, Tichy 1985). However, they only apply these notions, including product diversification, concentration and the establishment of state monopolies, to the final stage they propose. In other words, they only partially re-embody a disembodied theory advanced by Vernon.

Other popularised versions of the product-cycle model completely neglect the local enterprise dimension of the internationalisation of production and address themselves only to the overseas search by multinational corporations for new markets and cheap labour. Combined with the proposed addition of an eclipse phase, this exclusive concentration on the operation of multinationals illustrates the distortion and further simplification of an already simplified model. The product-cycle model and its variants do, in fact, perpetuate the idea that the ownership dimension of enterprise can be adequately represented by a simple binary variable of 100 per cent or zero that can be applied to either indigenous or foreign ownership. In reality, ownership and control is a much more complex matter. It involves joint venture agreements, local equity participation, local management participation, the nature of contractual arrangements between pairs of enterprises involved in trading relationships, and host and home government policies on industrialisation and investment. There is a wide range of corporate strategies that enterprises (and not just multinationals) can follow just as there is a wide range of roles that the separate sub-units of an enterprise can play within the enterprise as a whole (mother/daughter relationships, autonomous branches, product and geographical divisions, cash cows, or sub-units linked through matrix organisation). The product-cycle model has neither been constructed nor modified to accommodate this complexity. Instead the heavy and homogenising hand of economic rationality has been imposed.

Invention and innovation

The foundation of the product-cycle model is technological change and the aging of new technologies. However, the treatment of invention and innovation within the model is wholly unrealistic. It is implicit within the product-cycle model that inventions are introduced to the market in their final form. Empirical research would suggest that such an assertion is almost totally erroneous. Instead, the initial introduction of a product to the market is followed by progressive modification and improvement and invention in itself is most frequently incremental rather than revolutionary. The product that begins a cycle may bear little resemblance to the product that ends it, if an end can be discerned at all.

It has been contended by Gold (1981) that the development of a new product can not be divorced from the development of its market. The sequence begins with the introduction of a new product for which there is a given volume of potential demand. That market is progressively satisfied but with modification of the product a new market segment may be opened up. As such, the typical 'S-shaped' curve usually used to depict the progressive adoption of an innovation and the output curve of a new product as it might pass through the product-cycle, should be more properly seen as being paralleled by stepped curves depicting the progressive expansion of the potential market for the product as it is systematically improved.

By not making allowance for market expansion in parallel with incremental product improvement, it can be suggested that the product-cycle might be an artificial long-term construct inappropriately superimposed on a series of short-term events (Taylor 1986). If this is true, the product-cycle becomes a pseudo macro-cycle that can be cut at any stage. Since few products can be identified that have completed the cycle, this conclusion is not without weight.

Product homogenisation

A third limitation of the simplified product-cycle model is the way in which products are, by implication, homogenised. A car is a car irrespective of make or mark or increasingly sophisticated internal componentry. Micro-chips are micro-chips irrespective of the incompatible families of hardware they are used to produce. A sewing machine is a sewing machine and a television is a television and so on. Furthermore, no account is taken of the attempts that companies make to differentiate their products. The same generic drugs are marketed as vastly different products by pharmaceutical companies. One detergent manufacturer may market vitually the same product under different rand names to target specific segments of even the same national market. Commercial reality appears to hinge on product differentiation while the product-cycle model is built on assumed homogenisation.

What is more, while the model also deals with products singly and in isolation, commercial reality would see multiple products and services being combined within individual business enterprises, especially multinational corporations. As such, the fortunes of individual products and services are interdependent within the enterprises that produce or provide them. There are, therefore, significant commercial reasons reflecting corporate organisation and reorganisation that might lead to the truncation of a particular product cycle. The disembodied nature of the original product-cycle model does not allow these issues to be canvassed. It is also significant that in proposing an 'eclipse phase' to complete the model, it is the enterprise and organisational characteristics of the economic environment that are thrown to the fore.

It has also been contended that the product-cycle refers only to specific types of good, and Tichy (1985) has rehersed this proposition in detail. Unfortunately, this modification of the basic product-cycle model becomes trapped in circular reasoning. For Tichy (1985) product-cycle goods need agglomeration for their birth and early development, but not all goods created in agglomerations are product-cycle goods. Then, in their later stages of development product-cycle goods emigrate to more peripheral locations. Some goods may remain in the agglomerations of industrialised countries for a very long time and, indeed, Vernon (1966) went to some length to emphasise the range of forces that might delay the transfer of production from the home country. So, if a good follows the product-cycle it is a product-cycle good. If the cycle is truncated then it is not a product-cycle good. Anything can be proved by such an approach.

Location specific advantages

At the foundation of the product-cycle model is the location specific advantages for the creation of new products that is thought to exist in industrial agglomerations in industrialised countries. This argument is based on the existence and availability of local external economies in such locations. This is an argument which is frequently rehearsed (for example Townroe and Roberts, 1980) but rarely proved. No conceptual advance has been made in the study of local external economies since they were fully rehearsed in the work of Florence (1948). Their existence is a classic case of argument by assertion, and it is intriguing that the archetypical industrial agglomeration, the West Midlands conurbation in the United Kingdom has, for the best part of a decade, been in deep recession despite the much-vaunted resilience of its industrial structure resulting from the presence of local external economies.

However, the basic product-cycle model ignores the possible influence of location specific advantages in phases of development following the introduction of a new product. There is a pre-occupation in the model with the movement of the production of standardised goods to LDC's. It is implicitly assumed in the model that inputs other than cheap labour available in the LDC's are not only uniformly priced but also ubiquitous. There are two problems with such a contention. First, there is a large literature which suggests a wide range of location specific advantages, in addition to the availability of cheap labour, that might influence corporate decision-making and the orientation of internationalised investment and production. This literature has been brought together by Dunning (1981) in his eclectic theory of multinational investment. Second, the developed and industrialised countries are still the main recipients of foreign direct investment rather than the LDC's. The magnitude of these cross-flows are so great that intra-industry direct foreign investment has now been recognised as a discrete subject for research (Erdilek 1985).

This problem with the product-cycle model concerning its partial recognition of the location specific advantages that might influence the spatial aspect of investment decision-making is, indeed, symptomatic of a deeper malaise within the model. Within the model, the vast majority of the external relations of individual enterprises are ignored. Markets and marketing are ignored. Unequal trading relationships between pairs of businesses goes unrecognised. Contractual aspects of interaction

Product-cycle model

are not incorporated and the difference between intra-firm and inter-firm trade have no part to play in the model. In short, not only does the product-cycle neglect the enterprise context within which the process of product evolution must necessarily take place, it also treats the nature of the wider economic environment within which those enterprises must function in a wholly unrealistic fashion.

Scale, labour and less developed countries

Fundamental to this unrealistic treatment of the enterprise and its environment in the product-cycle model is the significance attached to the search for economies of scale and the need for cheap labour in the production of mature, standardised products. The assumed inevitability of the manufacture of standardised products being relocated in labour-intensive, mass production branch plants in less developed countries must be seriously questioned. As Vernon (1966) recognised, production in LDCs has very specific requirements. The product must be straightforward to make, it must have a high unit value to absorb freight costs, labour must be a principal input in its manufacture, and the product must have a high price elasticity. Furthermore, mass production need not be labour intensive. It can just as easily be capital intensive. Also computer numerical control (CNC) and computer aided manufacturing (CAM) have revolutionised batch rather than mass production by reducing set-up costs rather than line production costs. Standardisation and mass production need not be the inevitable final phase in the making of a given product. Associated with this nexus of issues that undermine the basic propositions of the product-cycle model is the fact that capital may not be as mobile as the model might suggest. The inertia of capital investment is too easily forgotten and the assumption of perfect capital mobility can be too glibbly applied.

Here again, reality is at odds with the assumptions of the product-cycle model. The proponents of the model have, in effect, inappropriately extended and universalised a set of circumstances in which one or a few products at a particular point in time might benefit in economic terms from mass production in cheap labour LDCs.

Standardised products

Within the product-cycle model there is also an assumed historical tendency towards stability and the standardisation of products together with their associated production processes and markets. However, through empirical analyses Walker (1979) has cast serious doubt on the validity of this set of inter-related propositions. He uses the case study of the textile machinery

industry and analyses by Rothwell (1976) to show that continuous
innovation and the 'out-innovating' of competitors is necessary
for firms to survive in some industries. He concludes that the
situation in the textile machinery industry:

"...bears little resemblence to the Hirsch/Vernon product-
cycle formulation. ... technology is changing rapidly, is
dependent throughout on external economies, and competitiveness can only be procured by sustained innovative effort and
increase in productivity; capital intensity tends to be
consistently high...; industry structure tends to be
oligopolistic, and financial resources, know-how, and a
supporting industrial infrastructure are required for
successful entry; critical human inputs are scientific and
engineering, managerial (including marketing and after-sales
services) and skilled operatives throughout; and demand
structure tends to be characterised by relative price
inelasticity." (Walker 1979, 25)

He extends this conclusion by suggesting that the same situation
exists in a wide range of 'systems technology' industries
producing aircraft, heavy electrical machinery, computers,
nuclear energy, specialist machine tools and scientific
instruments. Continuous technological change and instability also
characterises industries where governments have intervened for
military or other reasons to speed technological change.

Walker (1979) goes on to reinforce this point using
information from the consumer electronics industry. In television
manufacture 'component insertion' makes up 75 per cent of labour
costs. The pace of technological change in the electronic
components industry has left television assemblers with the need
for flexible production processes, giving a locational advantage
to low labour cost countries. Miniaturisation has reduced the
number of components in a television and now computer-guided
component insertion systems are being developed in the USA and
Japan. The suggestion is, therefore, that the comparative
advantage in television production will once more return to
capital-abundant developed countries. The main point made in this
example is, however, that production can remain labour-intensive
because of technological instability, not stability.

Monopoly power

The question of product standardisation as it is conceived in the
product-cycle model is also closely related to the role of
monopoly power. It is implicitly assumed in the model that
monopoly control of a product declines as the product ages and as
information and know-how about that product disseminates
throughout the economic community. What analyses by researchers
like Walker (1979) demonstrate is that instability and continuous

product improvement maintain monopoly power. The development of a new product by a particular business enterprise may stimulate a competitor to develop a substitute employing similar or totally different technology. But, the competitor can never quite catch up for the original innovator will simultaneously work on a second generation of his product. What develops, therefore, is a lagged response to innovation between groups of competing business organisations, with a process of leap-frogging being encouraged to maintain monopoly power. Such a process can also be an alternative to Coasian internalisation whereby an enterprise retains control of proprietary knowledge and extracts monopoly rents from it by retaining information within the subunits of the frequently multinational organisation.

In relation to the exercise and maintenance of monopoly power, therefore, the product-cycle model tends to misinterprete the way in which information is created and disseminated within and between firms. This stems from the inadequate conceptualisation of intra- and inter-organisational processes within the model, even in its extended forms.

Process and product innovation

The product-cycle model is also somewhat ambiguous in the way it treats process as opposed to product innovations, and this problem has been particularly well demonstrated by Walker (1979) in the context of the international chemicals industry. With the introduction of a new chemical product it would be expected that a manufacturer would be able to enjoy a long period of monopoly control as the development of such products is expensive and risky and patents on them are hard to circumvent. As the product matures and enters the final stages of the cycle, competition increases as does overseas production, comparative advantage depends on production costs and demand is price-elastic. At this stage, the product-cycle model would predict that production processes would also be at an advanced stage of standardisation (Walker 1979).

However, the history of the chemicals industry would suggest that such process standardisation is far from the case. Instead, as in the case of acrylonitrile for example, process technologies change radically bringing massive savings from mass production and slashing unit production costs. This tendency has had massive effects in the chemicals industry. Firms look to large home markets as secure bases, establish cartels to control price competition and stagger investments to avoid over-supply and price competition. What is more, imports from LDCs are usually small and only temporary. In short, process innovation can undermine apparent product stability. This is an important caveat to attach to the product-cycle model: it is not just what is made (the product) that is relevant, it is also the way that it is made that influences stability, monopoly power, inter- and intraorganisational relationships and the rate and direction of technological change.

Cycles upon cycles

It is implicit within the product-cycle model that, as a product develops, so the market expands to absorb all that is produced. Clearly, therefore, production is envisaged as driving the market and not vice versa, and a market cycle is inferred which parallels the product-cycle itself. Such a proposition is not defensible. The product-cycle may be terminated by market pressures and Sayer (1985) has, in fact, bemoaned the fact that few products appear to complete the postulated cycle.

Taylor (1986) has demonstrated the cutting of the product-cycle by a discordant market cycle in the context of the manufacture of electronic micro chips from silicon and germanium. This work also suggested that there is also a third cycle, an enterprise cycle, which interacts with both the product-cycle and the market cycle. As such, the preoccupation with just one cycle relating to the invention and progressive obsolescence of products neglects the significant point and counterpoint of market-, enterprise- and product-cycles that is the hall mark of reality. Without consideration of this interplay of cycles, the product-cycle must necessarily remain technologically deterministic.

The Product-Cycle Model and Regional Uneven Development

The simplicity of the product-cycle model has led to its ready incorporation into a range of studies in economic geography including analyses of spatial restructuring, technological change, the location of R&D activities, the creation and growth of firms, high technology and technology park planning (see comments in Taylor 1986). These studies share four principal characteristics in their use of the product-cycle model. First, the model has been accepted uncritically. The basic product-cycle model has not been questionned in any systematic way and not even the limitations to its applicability recognised by Vernon (1966) (for example, the need for cheap labour in production functions) have been accommodated. Instead, proponents in geography have caricatured, extended and universalised the model (see Sayer 1985).

Second, the users of the product-cycle model in geography have been concerned only with its spatial dimension, and the rationale that it provides for locational shifts, relocation and regional uneven development. This fetishising of space has occurred despite the warning by Thomas (1980) that the spatial dimension of the product-cycle model is less than adequately developed. Beginning with the lead given by Norton and Rees

(1979), an inevitability has been ascribed to the locational shifts associated with technological change. In effect, technological determinism has been bolstered by spatial determinism to create an explanatory straight-jacket which is now primarily applied to high technology industry and assessments of its regional employment generating potential.

Third, in fetishising space, the scale of analysis has been reduced from the international to the inter-regional. These contexts are functionally very different, with regional systems being extremely open and national and international systems being relatively closed. This translation in geography has been made uncritically.

Fourth, the disembodiment of technological processes from their enterprise context is perpetuated, despite the greater urgency and relevance of this context at the sub-national, regional and local scales with which geography has traditionally dealt. Through this approach, 'spatial' processes are given free rein with the effect that 'regional life-cycles' can be conjectured (Rees 1979), unhampered by the awkward intrusion of the processes of investment decision-making in management coalitions.

The combination of these characteristics of the product-cycle model as it has been imported into and used in geography is that a simple linear model has been created which links technological change, new firm formation and spatial adjustment. In this model, based on the product-cycle concept, R&D expenditure is held to create invention and innovation at R&D locations which, in turn determine the location of new industry. This sequence is taken to 'explain' high technology concentrations in such localities as Silicon Valley and Cambridge, and leads quite naturally to what Taylor (1986) has called 'cargo cult' planning to create economic growth and new employment opportunities in industry parks, technology parks or even in the UK's failed enterprise zones. The tenet of 'cargo cult' planning is to replicate the physical circumstances of an existing high technology location (housing, freeways, parkland surroundings, golf courses, good schooling and so on) because then industry will materialise and supply jobs. The lack of reality in these prescriptions is shown by the North Ryde technology park in Sydney, Australia. Certainly, Japanese high technology electronics firms are to be found there, but only their distribution warehouses! What has been forgotten by neglecting the enterprise context is that all firms, and especially large corporations, consist of separate sub-units arranged hierarchically that perform more or less vital functions within that enterprise. Therefore, in a functional, and not necessarily spatial sense, a firm has a core and a periphery of elements more or less central to its overall operations. The majority of geographical research that has used the product-cycle framework has substituted fetishised space for the complexity of investment decision-making and resource allocation that goes on within firms while at the same time maintaining the technological determinism of the original model.

Conclusion

The product-cycle model is an elegant but limited explanatory construct which has been inappropriately universalised by its proponents in a way not necessarily intended by its originators. The foundation of the model is technological determinism. In this chapter, nine areas of conceptual limitation inherent in the model have been outlined and elaborated. They relate to assumptions about enterprise production functions, ownership, the exercise of monopoly power, the processes of invention and innovation, product inter-relationships, standardisation and stability, economies of scale, the interplay of process and product technology and the nature of business environments.

In combination, these areas of criticism show that the deficiencies of the product-cycle model arise from its failure to put technological change and product development into an adequate enterprise context. The use of the model in geography has exacerbated this problem by crudely caricaturing the spatial elements of the original model. By implication, therefore, only through the development of an adequate conceptualisation of the enterprise, its structure, functioning and interaction with other enterprises and organisations, can the ambiguities and the conceptual limitations of the product-cycle model be overcome. The development of a conceptualisation of enterprise is an urgent task confronting economic geography as a whole.

References

Auty, R.M. (1984) 'The Product Life-Cycle and the Location of the Global Petrochemical Industry After the Second Oil Shock' Economic Geography, 60 (4), 325-38

Boudeville, J.R. (1966) Problems of Regional Economic Planning, Edinburgh University Press, Edinburgh

Dunning, J.H. (1981) International Production and the Multinational Enterprise, George Allen and Unwin, London

Caves, R.E. (1982) Multinational enterprise and Economic Analysis, Cambridge University Press, Cambridge

Coase, R.H. (1937) 'The nature of the Firm' Economica, 4, 386-405

Erdilek, A. (ed) (1985) Multinationals as Mutual Invaders; Intra Industry Direct Foreign Investment, Croom Helm, London

Florence, P.S. (1948) Investment, Location and Size of Plant, Cambridge University Press, Cambridge

Gertler, M. (1985) 'Regional Capital Theory' Progress in Human Geography, 8(1), 50-81

Gold, B. (1981) 'Technological Diffusion in Industry: Research Needs and Shortcomings' Journal of Industrial Economics, 29 (3), 247-69

Hall, P. (1985) 'The Geography of High Technology: An Anglo-American comparison' Paper presented to the conference on Innovation, Change and Spatial Impacts, Melbourne, Australia, August 1985

Hirsch, S. (1965) 'The United States Electronics Industry in International Trade' National Institute Economic Review

Hirsch, S. (1967) Location of Industry and International Competitiveness, Clarendon Press, Oxford

Norton, R.D. and J. Rees (1979) 'The Product Cycle and the Spatial Decentralisation of American Manufacturing' Regional Studies, 13, 141-151

Oakey, R.P. (1984) 'Innovation and Regional Growth in Small High Technology Firms: Evidence from Britain and the USA' Regional Studies, 18, 237-251

Parry, T.G. (1975) 'The Product Cycle and International Production: UK Pharmaceuticals' Journal of Industrial Economics, 23, 21-27

Perroux, F. (1950) 'Economic Space, Theory and Applications', Quarterly Journal of Economics, 64, 89-104

Rees, J. (1979) 'Technological Change and Regional Shift in American Manufacturing' Professional Geographer, 31, 45-54

Rothwell, R. (1976) 'Innovation in Textiles Machinery: Some Significant Factors in Success and Failure' Science Policy Research Unit, Occasional Paper No. 2.

Sayer, R.A. (1985) 'Industry and Space: A Sympathetic Critique of Radical Research' Society and Space, 3(1), 3-29

Taylor, M.J. (1984) 'Industrial Geography' Progress in Human Geography, ((2), 263-274

Taylor, M.J. (1985) 'Industrial Geography' Progress in Human Geography, 9(3), 432-442

Taylor, M.J. (1986) 'The Product-Cycle Model: A Critique 'Environment and Planning A, 18 (forthcoming)

Thomas, M.D. (1980) 'Explanatory Frameworks for Growth and Change in Multiregional Firms' Economic Geography, 56, 1-17

Tichy, G. (1985) 'Is the Product Cycle Obsolete?' Research Memorandum 8502, Department of Economics, University of Graz, Austria

Townroe, P.M. and N.J. Roberts (1980) Local-External Economies for British Manufacturing Industry, Gower, Farnborough

Vernon, R. (1960) Metropolis 1985: An Interpretation of the Findings of the New York Metropolitan Region Study, Harvard University Press, Cambridge, Mass.

Vernon, R. (1966) 'International Investment and International Trade in the Product Cycle' Quarterly Journal of Economics, 80, 190-207

Vernon, R. (1979) 'The Product Cycle Hypothesis in a New International Environment' Oxford Bulletin of Economics and Statistics, 41, 225-267

Product-cycle model

Walker, W.B. (1979) <u>Industrial Innovation and International Trading Performance</u>, Jai Press, Greenwich, Conn.

Williamson, O.E. (1975) <u>Markets and Hierarchies: Analysis and Antitrust Implications</u>, Free Press, New York

Williamson O.E. (1981) 'The Modern Corporation: Origins, Evolution, Attributes' <u>Journal of Economic Literature</u>, <u>19</u>, 1537-68

CHAPTER SIX

TECHNOLOGICAL CHANGE AND SPACE DEMAND IN INDUSTRIAL PLANTS

Dr. R. Grotz, Professor Department of Geography, University of Bonn[1]

Problem

In looking at the growing space demand of existing firms we are dealing with one of the most acute and distressing problems of old industrialised areas. Surveys on partly and totally relocated firms in West Germany reveal that, even in the seventies, the lack of space and buildings was the most important factor in relocation. The destination of expanding firms is well-known: most are just moving to the outskirts of their agglomeration area where they take up areas of open land.

During the last 10 to 15 years, metal processing technologies experienced such a rapid and fundamental change that their importance is now compared with the introduction of assembly lines. The change arises from the installation of electronic elements into machine tools and innovative process engineering. This produces a new generation of machines with improved qualities and capacities. The spatial consequences of such changes for sites and plants are discussed here.

Experience from the past is that an increase in production and/or installation of a new type of machinery very often also means a higher space demand. Because of restricted sites, enterprises could not meet this demand, and so many established branch plants or were even forced to relocate the whole firm. The question is whether or not even in a slowly growing economy the trend of increasing space demand for production purposes will continue.

Empirical Study

The answer to this question is rather complex, because it is affected by many factors, some of which are contradictory. This chapter describes the impact of technological change on space demand in metal industries and gives some general views on technological trends. Results ensue from a study in the Stuttgart area, which is marked by both its diversified industry producing investment goods and its lack of industrial land within the agglomeration area. The field survey was conducted by Harald Koehler. He visited some 60 small and middle-sized firms (up to 1,000 persons employed), which had recently installed new types of machines. In 20 firms it was possible to research the former situation and thereby compare shop layouts, production figures, processing times, etc. In this chapter only the main features and the results of the study are presented (for details Grotz and Koehler 1984). In addition, some experts were questioned in order to supply further information on aspects of factory planning, material flows and construction of machines.

The type of machines considered here are cutting machine tools. They can shape metal by turning, drilling, milling,

Figure 1: CNC-machining centre.

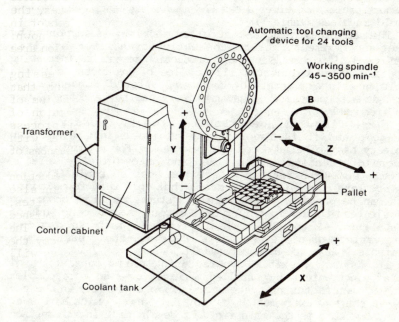

Source: Oerlikon

grinding, and honing. These machines can be operated by manual or electro-mechanical controls. If they are equipped with a simple electronic control they are called NC-machines (NC= numerical control), or if they use a more advanced system they are referred to as CNC-machines (CNC= computerised numerical control). There are further specialised and more sophisticated control systems (AC= adaptive control, DNC= direct numerical control), but in this chapter all are summarised under the expression 'NC/CNC' (Fig. 1).

Results

Advantages of the new generation of machines are manifold: they work faster and more precisely; once programmed, they can produce with little human supervision; and due to their automatic tool changing capacities they are able to execute several procedures in one place. Although their price is about double of that of a conventional machine, costs per unit output are lower, given a reasonable utilisation. Because of its multi-purpose character and short processing times a new CNC-machining centre can replace 2-5 old machines (Table 1).

The superiority of NC-controlled machines is supported by market developments. Up to the early seventies many firms experienced slow changes in their production programs. Increasing competition led to shorter life cycles of products and, as a result, a greater variety of goods was offered. Consequently, batch sizes became smaller and machinery had to be changed over more frequently.

These circumstances require more flexibility in the workshop, and it is important to minimise idle time when machinery has to be adjusted for new tasks. For producing great numbers of identical parts fixed and automatically working systems (e.g. transfer lines) are the most suitable (Fig. 2). They are less flexible, which means that only very few similar parts can be produced per machining system. But large economies of scale lead to low costs per unit output. At the other end of the scale are NC- and CNC- machining centres, which can work on a great variety of different parts. However, their use is only profitable when batch sizes are small.

In future, flexible manufacturing cells consisting of several electronically controlled machines or even more complex flexible manufacturing systems are supposed to fill in the gap between the two extremes. Up to now only a few units have been set up, but this is the main field of development in enterprises producing machine tools. The new systems will have to work economically within a broad range of batch sizes and need to be able to handle a big number of different parts.

Flexibility in this context means production of a great variety of parts which can be manufactured in any given sequence without long interruptions. The more complicated the shape of a

Table 1: Product of a machine component, batch size: 10

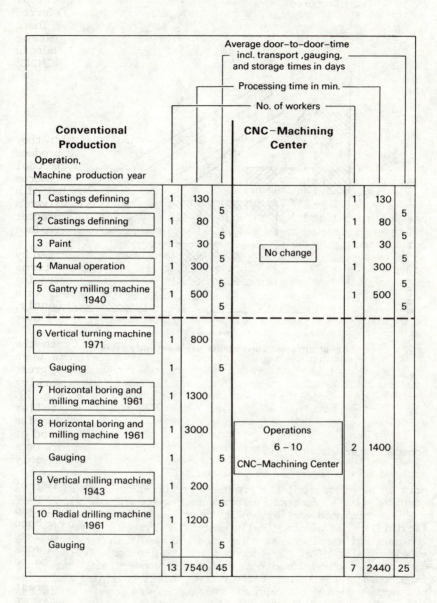

Conventional Production — Operation, Machine production year	No. of workers	Processing time in min.	Average door-to-door-time incl. transport, gauging, and storage times in days	CNC–Machining Center	No. of workers	Processing time in min.	Average door-to-door-time incl. transport, gauging, and storage times in days
1 Castings definning	1	130			1	130	
			5				5
2 Castings definning	1	80			1	80	
			5				5
3 Paint	1	30		No change	1	30	
			5				5
4 Manual operation	1	300			1	300	
			5				5
5 Gantry milling machine 1940	1	500			1	500	
			5				5
6 Vertical turning machine 1971	1	800					
Gauging	1		5				
7 Horizontal boring and milling machine 1961	1	1300					
8 Horizontal boring and milling machine 1961	1	3000		Operations 6 – 10 CNC–Machining Center	2	1400	
Gauging	1		5				
9 Vertical milling machine 1943	1	200					
			5				
10 Radial drilling machine 1961	1	1200					
Gauging	1		5				
	13	7540	45		7	2440	25

Source: a Stuttgart machine tool firm.

Figure 2: Performance of production concepts in dependence of batch size and number of different parts to be manufactured.

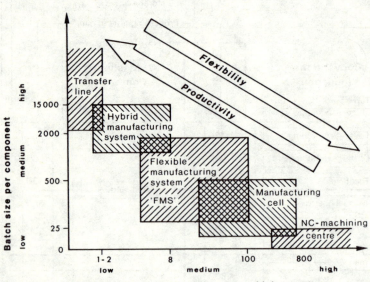

Source: after Klahorst (1978).

part to be processed, the greater are the advantages of NC/CNC compared with a conventional production system (Table 2).

As mentioned earlier, due to improved performance and higher flexibility, an electronically controlled system can replace 2-5 conventional machines. But this does not result in a reduction to a half or a fifth of the former space requirements. NC- and CNC-controlled machines need more space than conventional ones. This is because of their larger floor area, complete lining for noise protection, leading devices, other peripheral devices, and the extra room needed for a programmer, a second set of tools, a place for tool presetting devices, an so on (Fig. 3).

Table 2: Production time and space requirements of conventional and CNC-turning machines.

Space requirements	"CNC" 22,5 m^2		Conventional 12,5 m^2
Part	**Difficulty for production**	**Ratio of time** CNC:Conventional	**CNC space requirements** compared with conventional machines of same output
Turned disk	simple	1:1,5	extra demand 3,75 m^2
Tool holder	normal	1:2	savings 2,5 m^2
Adapter for machine tool	difficult	1:3	savings 15 m^2

Source: a Stuttgart machine tool firm.

Exact comparisons of the space situation in 20 plants before and after replacement of machinery revealed the following: in half the cases there was no significant change in space demand ($\pm$ 5m^2 per NC/CNC-machine). Two firms expanded their production area. But in eight cases considerable savings of room could be achieved (5-30 m^2 per machine) in addition to the achievement of greater efficiency and other advantages listed above. These eight firms are also using other peripheral equipment (e.g. automatic transport and loading devices), and they had altered organisation structures. Enterprises experiencing no space advantages did not manage to adjust their workshops organisation to the faster performance of the new machines: that is, additional storage area is used within the shop as a buffer between different operations (Fig. 3).

These different results can be explained by some basic developments: to achieve more flexibility and at the same time to lower costs per unit output are the most important aims of production planning. Both can only be achieved by automation. Before automation of a process can be introduced successfully,

Figure 3. Models of a conventional production line and a NC-machining centre, showing layouts, material flows and intermediate storage areas

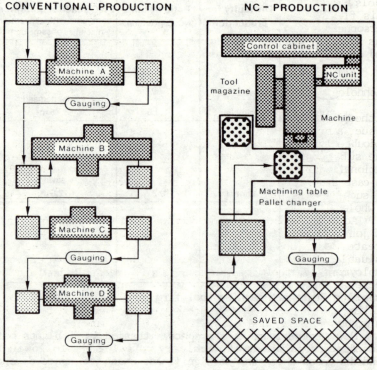

CONVENTIONAL PRODUCTION

Machine A

Gauging

Machine B

Machine C

Gauging

Machine D

Gauging

Operation time. 4 weeks

NC - PRODUCTION

Control cabinet

Tool magazine

NC unit

Machine

Machining table
Pallet changer

Gauging

SAVED SPACE

Operation time 3 days

the material flow has to be reorganised. For this, containerisation of the workpieces is necessary, which will need more space than before. Equally, semi-automation demands more room because preceding and ensuing conventional procedures can hardly be coordinated with the faster mode of operation of the new generation of machines. Maximum cost advantages can only be achieved by a new production organisation adapted to the machines. Utilisation time increases, material flow is accelerated and storage can be reduced.

Integrated production systems consisting of several NC/CNC-machining centres plus superior control will reduce organisation problems even for small batch sizes. Compared with single NC-machines these systems require more room, but this is again outweighed by greater efficiency. The example of a Messerschmitt-Bölkow-Blohm plant in Augsburg reveals a number of additional advantages. In this plant not only are all metal-cutting procedures computerised but also provision of materials and tools, waste discharge and stock control are optimised. Using this flexible control system, savings in processing time and of personnel were 44 per cent each, while reduction of floor space was 39 per cent. Total door-to-door-time was reduced by a quarter and costs were lowered by about the same percentage.

Technological Change and Labour

Although in this contribution the aspect of labour is not at issue, a brief outline of the impacts on labour shall be given. Undoubtedly, electronically controlled machines dismiss workers. The substitution of labour by capital is the main effect of rationalisation. How many jobs are lost can only be established in case studies where the conditions are well known. The results of such case studies can't simply be projected for industries as a whole (Dostal 1985).

Actually, there are some developments counteracting the loss of jobs. Though in future increase in output only rarely will create more jobs, qualitative growth by innovative and more valuable products are more likely to have positive effects on employment. Furthermore, in many highly industrialised nations a decrease in working hours per year and employee is being experienced, which also leads to more jobs. In this way, job-killing effects of modern technologies are weakened by increases in product quality and reduction of working time.

New technologies also brought about considerable changes in demands on qualifications. The case study shows two contrasting tendencies:
1. If programming, correction of programs, organisation of production and perhaps small repairs belong to the responsibilities on the shop floor, very specific knowledge is required. Only highly skilled workers can perform these tasks.
2. If programming, planning, tool setting and repairs are handed over to specialists, skilled workers are hardly needed at the machines. Under these conditions they can be operated by semi-skilled workers.

Both tendencies are closely related to applied technologies, batch sizes and factory sizes. The older NC-machines supported the division of labour. The newer CNC-machines can be operated and programmed more easily; therefore they are only used for small and middle batch sizes. Once again it is possible to delegate more responsibility to the shop floor (Kern and Schumann 1984, Sorge 1985). This is mainly done by smaller enterprises. Bigger enterprises, which have always preferred the division of labour, mostly retain their old structures. The second of the above mentioned tendencies dominates, as the use of CNC-technology in the FRG was more common in large firms than in small ones (82 per cent in firms with more than 500 persons employed, 24 per cent in single plant enterprises with less than 100 persons employed; Ewers and Kleine 1983). This means that several new professions of high qualifications are necessary (electronic engineer, programmer, workshop planner, etc.), but at the same time qualification standards at the machines become lower. In the case study seven out of 20 plants revealed they had acquired NC/CNC-machines mainly because of a lack of skilled workers.

Trends in Technical Developments

Cost reduction is the leading force behind technological change, but many developments are fostering a trend to a higher production capacity per unit area:
Storage areas: in new high-rise buildings former storage areas are reduced by up to 60 per cent.
Machinery: the latest CNC-machining centres exhibit a more compact design than the first types of NC-machinery. Microprocessors and computers shrink control cabinets to the size of small boxes, and also the expense of programming is reduced. At the same time reliability is improved, and the machines are easier to maintain and to repair. Hence more and more old, stand-by machines became unnecessary. They formerly cramped many workshops and were preserved for use in case of breakdowns.
Product design: since new products contain fewer materials and parts the amount of mechanical tooling is decreasing. The transition from electro-mechanical components to electronic elements cuts drastically the number of parts needed for a final product. In a teleprinter one microprocessor replaced 936 electrical and mechanical components.
Process engineering: a few years ago new methods of process engineering appeared which can reduce time and space consuming operations. Improved injection moulding of plastic materials, precision die casting and forging of metals, or electrical erosion processes replace former assembly techniques of complicated parts with one operation.
Assembly: so far attention has been mainly directed towards automation of the machining part of production. Usually the automation level of assembly lines is much lower. To upgrade this is arguably the major task of engineers in this decade.

Although robots have been used for 20 years, there were only about 7,000 in the Federal Republic of Germany by the end of 1984. Their main fields have been the handling of workpieces and tools. They often did not replace human labour, but substituted other machines that had been operated by severe physical exertion or could bring health damages. Robots of the second generation are equipped with optical and touching sensors, by which they can recognise, differentiate and control. These abilities will enlarge their use to the labour intensive assembly area. Even now a trend can be recognised to construct new products suitable for automatic fabrication, assembly and control.

Robots need a security zone within their reach; therefore they need more room than a worker in the same position. Savings in space demand are expected, though, when robots operate more than one machine and adjustment to special jobs is improved.

In future, increased flexibility of production and assembly lines will allow the manufacture of complete product families at a similar cheap cost as only specialised systems did up to now. Where several perhaps partly utilised production lines had existed side by side or lenghty adjustment procedures had been necessary, often few lines will do in future, which can be adjusted quickly due to easy programming.

As long as there are only few automatic assembly lines, flexibility is increased by desintegration of linear job organisation into assembly islands. This structure allows for a more versatile employment of workers; however, it needs more space. As this shows, there are also trends for an increase in space demand. Among others, steps taken towards humanisation of the work environment or the installation of social facilities are included.

Consequences for Locations and Sites

1. As both production volumes and process engineering are changing rapidly, it is very difficult quantitatively to predict land demand for industrial use. But one point should be stressed. Urban and regional planners as well as manufacturing engineers are using industry specific figures for calculating industrial land demand (e.g. m^2 per person employed). These figures gained from experience in the past are published in textbooks, and have been printed unchanged over decades. In applying these figures, the characteristics of antiquated technologies are projected into future situations. New techniques are marked by both a higher production intensity per unit space and a significantly higher labour productivity. Therefore, given figures for space demand have to be applied with care.
2. To save money, modern factories are built in a compact style. As a result, space demand is less, distances of transport become shorter and costs for heating and cooling are reduced. Even multi-storeyed buildings are economical if the material flow per time unit is small (transport of few and/or small resp. light parts), which is typical of many high-tech products e.g. in the electronics, precision engineering and optical industries. However, the planning of production

103

processes and material flows is more difficult, since two or more levels have to be co-ordinated. Therefore reorganisation due to new requirements can be more difficult than in a single-storey building. Advantages and disadvantages have to be considered carefully for every case.

3. In respect of crowded industrial areas two questions will gain more and more importance. How can established plants modernise production structures, and how can they achieve more flexibility within their given site? New types of machine tools and manufacturing engineering enable firms to respond quickly to market requirements. Even on restricted sites there is a chance to obtain more flexibility, efficiency and capacity. This favours inertia in locational behaviour.

4. Space costs are less important than many other cost factors, for they amount to only 1-2 per cent in the price per unit output. This means that the space demand of production is of major relevance only if the firm is rather cramped or if land prices are very high. Savings in space are only a by-product of the general aim to reduce costs.

Until now, only few industrialists have noticed the space saving qualities of the new machinery. Even professional plant planners are hardly aware of them. Such aspects as, for example, higher flexibility, clear material flows and a strict distinction of functions rather than an optimal utilisation of space are primarily considered. Lack of space usually leads to extensions or new buildings (Table 3). As this is often difficult or even impossible at the old site, relocations or production splits are common solutions of the problem. Neglect of the new space saving possibilities in the production could contribute to a continuing trend of relocations and new branch plants. Experience from the survey indicates that the use of new machinery and consequent adjustment of process flows and of new organisation patterns have already obviated the need for firms to relocate. In one case a nearby branch plant became unnecessary and was closed.

5. An aspect rather overlooked here is the rapid substitution of labour by capital. In order to make better use of investment more shifts per day are required. Two shifts have been introduced almost everywhere, but there have been attempts with a third shift manned by a reduced staff. In a plant of the Heidelberger Druckmaschinen AG each of the two shifts is followed by two hours of automatic production. As a consequence, machinery is used 20 hours per working day. Reliability of equipment still does not allow for full automation. Automatic production requires the storage of many workpieces before and after processing. In spite of this extra need for store-room and space for spare tools, tool setting devices, etc. production intensity per unit area will increase once more.

6. A number of empirical studies, some of which are published in Thwaites and Oakey (1985), supply information on the

Table 3: Solutions for an assumed lack of space in production area (20 firms, multiple response possible).

Possibilities	No. of answers
Multi shift production (already introduced in most cases)	1
Better production control	2
Removal of old machinery	2
New building	3
Subcontracting of components	3
Total change over to NC/CNC-technology	4
Expansion by an annex	10
Total	25

Source: field survey

adoption-nonadoption of process innovations and their spatial consequences. Naturally, all studies reveal major differences depending on the type of industry examined. Undoubtedly, early users of new technologies are large enterprises, often with their own R & D facilities. They can bear the risks of new technologies more easily, and their financial sources are better than those of single plant enterprises. Furthermore, big firms can use them for more variable tasks ensuring a minimum threshold. Investments with unprofitable economies of scale may become a problem for small enterprises.

A study on the diffusion of NC/CNC-technologies in the mechanical engineering industry of the FRG revealed no influence of urbanisation variables. But localisation variables turned out to be the main determinant of interregional variations (Ewers and Kleine 1983). As small, independent single plant enterprises are over-represented in less industrialised areas, acceptance of new technologies is lower there. In remote areas infrastructures, especially in the information sector, are less developed. The availability of technical information and advice can vary considerable from region to region. Small labour markets often do not offer the qualifications needed to adopt 'imported' process innovations. These structural deficits can only be compensated by a technology orientated economic and regional policy (Rothwell 1982, Sayer 1983).

7. In the past eight years, the net growth of the West German economy has always been considerably below the growth of technological productivity which is slightly more than 3 per cent at the moment. Theoretically, productivity could have even grown faster if the age of equipment had not steadily increased. Generally speaking, low growth rates in output, as they were experienced in the past, should not develop new demands of space. Empirically, it takes 20 to 40 years until

innovative technologies are widely established. The diffusion
of space saving NC/CNC-technologies among large enterprises is
almost completed. In many middle and small size firms,
however, old and disorderly production structures still have
to be adapted to new requirements. As a consequence, more
space may be needed in the next 10 years, until the savings of
smaller and more effective machines with optimised process
flows will dominate.

Note

1. The author would like to thank Harald Koehler for co-operation
   and his excellent field work, Dr. Hans-Jürgen Warnecke and
   Manfred Hofmann for discussions and helpful comments.

References

Borchard, K. (1983) 'Arbeitsstätten'. In G. Albers et al. (Eds)
Grundriss der Stadtplanung, Akademie für Raumforschung und
Landesplanung, Hannover, 199-208

Dähnert, H. (1982) 'Beispiele flexibler Fertigungskonzepte und -
systeme' Technisches Zentralblatt für praktische
Metallbearbeitung, 76, 42-56

Dostal, W. (1985) 'Neue Technologien als Job-Killer? Quantitative
und qualitative Auswirkungen neuer Technik auf den
Arbeitsmarkt' Der Bürger im Staat, 35, 95-99

Ewers, H.-J. and J. Kleine (1983) The Interregional Diffusion of
New Processes in the German Mechanical Engineering Industry.
Discussion Papers IIM/IP 83-2, Wissenschaftszentrum Berlin

Gibbs, D.C. and A. Edwards (1985) 'The Diffusion of New
Production Innovations in British Industry' In: Thwaites,
A.T. and R.P. Oakey (Eds) The Regional Economic Impact of
Technological Change. London, 132-163

Grotz, R. and H. Koehler (1984) 'Macht die technologische
Entwicklung neue Zweigwerke überflüssig? Überlegungen zum zu
künftigen Flächenbedarf der Industrie' In: Elsasser, H. and
D. Steiner (Eds) Räumliche Verflechtungen in der Wirtschaft,
Zürcher Geographische Schriften, 13, 77-93

Hottes, K. and H. Kersting (1977) 'Der industrielle Flächenbedarf
- Grundlagen und Messzahlen zu seiner Ermittlung' In:
Siedlungsverband Ruhrkohlenbezirke (Eds) Konzeption zur
Industrieansiedlung. Essen, 223-277

Kern, H. and M. Schumann (1984) Das Ende der Arbeitsteilung?
Rationalisierung in der industriellen Produktion:
Bestandsaufnahme, Trendbestimmung. München

Klahorst, T.H. (1978) 'Rationalisierung durch flexible
Fertigungssysteme' Werkstatt und Betrieb, 111, 578-582

Koehler, H. (1982) Auswirkungen technologischer Neuerungen in der

Fertigung auf den Flächenbedarf der Industrie, Geographisches Institut Universität Stuttgart (unpublished)

Ledergerber, A. (1980) 'Produktionstechnik in den neunziger Jahren' VDI-Bericht, 390, 16-22

Marandon, J.C. (1980) 'Die industrielle Flächenplanung im technologisch-organisatorischen Prozess der Industrieentwicklung' In: W. Gaebe and K. Hottes (Eds) Methoden und Feldforschung in der Industriegeographie, Mannheimer Geographische Arbeiten, 7, 53-68

Müller, J. (1981) 'Computergesteuerte Maschinen. Die Oekonomie automatisierter Produktionsprozesse und deren Auswirkungen auf die Arbeitsanforderungen' Campus Forschung, 228, Frankfurt, New York

Podolsky, J.P. (1975) Methodik der Ermittlung und Anwendung von Flächenkennzahlen für die Grobplanung von Fabrikanlagen, Diss. Thesis Fakultät für Maschinenwesen, Technische Universität Hannover

Rees, J., R. Briggs and D. Hicks (1985) 'New Technology in the United States Machinery Industry: Trends and Implications' In: Thwaites, A.T. and R.P. Oakey (Eds) The Regional Economic Impact of Technological Change, London, 164-194

Rothwell, R. (1982) 'The Role of Technology in Industrial Change: Implications for Regional Policy' Regional Studies, 16, 361-369

Sayer, A. (1983) 'Theoretical Problems in the Analysis of Technologial Change and Regional Development' In: Hamilton, F.E.I. and G.J.R. Linge (Eds) Spatial Analysis, Industry and the Industrial Environment, 3 - Regional Economies and Industrial Systems, Wiley Chichester, 59-73

Schliebe, K. (1982) 'Industrieansiedlungen' Forschungen zur Raumentwicklung, 11, Bundesforschungsanstalt für Landeskunde und Raumordnung, Bonn

Sorge, A. (1985) 'Auf dem Wege zur "automatisierten Fabrik"? Industriearbeit unter den Bedingungen neuer Technik' Der Bürger im Staat, 35, 86-89

Thwaites, A.T. and R.P. Oakey (Eds) (1985) The Regional Economic Impact of Technological Change, London

Warnecke, H.J. and W. Dangelmeier (1982) 'Materialflusskosten minimieren mit integrierten Systemen kann gebundenes Kapital senken' Maschinenmarkt, 88, 38-40

Warnecke, H.J. and K.G. Lederer (1979) 'Neue Arbeitsformen in der Produktion' VDI-Taschenbücher, 52, Düsseldorf

Warnecke, H.J. and R.E. Scheiber (1982) 'Flexible Arbeitsorganisation zur kurzfristigen Fertigungssteuerung' Werkstatttechnik - Zeitschrift für industrielle Fertigung, 72, 381-384

Warnecke, H.J., R. Steinhilper and W. Schuetz (1982) 'Flexibel automatisierte Teilefertigung in mittelständischen Unternehmen' VDI-Zeitschrift, 124, 611-619

CHAPTER SEVEN

TECHNOLOGY AND INDUSTRIAL CHANGE: THE ROLE OF THE EUROPEAN
COMMUNITY REGIONAL DEVELOPMENT POLICY

Prof. P.S.R.F. Mathijsen, Director General for Regional Policy,
European Commission, Brussels

Introduction

It is a privilege for someone who is responsible for the economic
development of the less favoured regions of the European
Community to be able to confront his views with experts in the
field of international geography. The topic chosen for the
Congress of the Commission on Industrial Change seems to me most
appropriate for the problems which are facing us in Europe
today.

Let me start by telling what objectives we are trying to
achieve by implementing a Community regional policy. There are,
as you know, presently ten, and soon there will be twelve Member
States in the EEC each one with its own regional development
policy and measures. The main task of the European Commission is
to coordinate those national endeavors and complement them with
its own efforts. The necessity for Community intervention in this
field seems obvious: the regional disparities are enormous and a
real 'common market' cannot operate under such conditions; the
very existence of the Community therefore requires that everything be done to allow the backward regions to develop their own
economic activities, because this, in turn, would allow those
Member States with the greatest regional problems to implement
macro-economic measures which are convergent with those of other
Member States and the Community as a whole.

The regional policy of the Community has over the past few
years followed four guidelines. In the first place it has
implemented a method of regional analysis which allows it to
follow the socio-economic evolution of all the regions of the EEC
and to classify them according to their needs and potentialities;
this classification is based on the classical main parameters GNP
per capita and structural unemployment and summarised in the
so-called 'synthetic index'.

In the second place the Commission tries to coordinate the
regional policies of the Member States not only where their
systems of incentives are concerned but also their other policy
measures; one of the principal tools for this harmonisation is

the regional development programme which each Member State is required to draft and submit to the Commission and the Regional Policy Committee: those pluri-annual programmes allow a comparison of the specific actions to be undertaken in the various regions of the Community and to avoid duplications and contradictions.

The third guideline is probably the most important one, it is referred to as the analysis of the regional impact of other policy measures. This action is based upon the principle that any economic measure can have an important effect on the development of any region. This fact is very often overlooked by those who are responsible for policy measures and it is therefore the task of regional policy to make sure the effects on the regions are analysed and taken into account when decisions regarding those other policies are taken. On the other hand regional policy measures can be implemented which will facilitate the application of those other policies or even compensate for their negative effects.

Finally there is, of course, the European Regional Development Fund (ERDF) through which the Commission gives financial support for investments in infrastructure, industry and tourism and for studies. The Fund's endowment has increased regularly over the years to reach some 3.5 billion ECU in 1986 after the enlargement with Portugal and Spain. Since the beginning of 1985 a new regulation defines the conditions under which the resources are to be committed; some of the new features concern the financing of pluri-annual and comprehensive programmes for given areas (instead of project-financing), the emphasis on small and medium-sized enterprises and the development of the endogenous potential of the regions. The latter aspect is of particular importance since it seeks to create economic activities in the needy regions through the development of existing potential and initiative rather than by 'importing' activity from outside. As a matter of fact this stimulation of endogenous potential has become the leitmotiv of regional policy of the EEC with an emphasis on productivity and competitivity. The latter of course can only be obtained through the application, in the regions, of the new technology and this is where the theme of this book comes in. I would like to discuss three aspects in connection with this theme and what I have just said about Community regional policy: industrial change and new technology in a Community perspective, the impact of industrial change and new technology on the regional development of the Community and finally the role of Community regional policy in regard to industrial change and new technology.

Industrial Change and New Technology in a Community Perspective

Industrial change

Industrial change has become a matter of increasing concern in the Community. Whilst total industrial employment continued to grow until the early seventies, there has been a turnaround since then and between 1970 and 1983 industrial employment decreased by over 17 per cent. This overall development has not been due to job losses in some sectors and growth in others. In fact, employment decreased in all branches, with the decline particularly high for raw material products (e.g. ore, coal, steel), and for labour intensive sectors such as textiles and clothing, footwear and leather. Projections for the period up to 1990 carried out by the Commission (in the Second Periodic Report on the social and economic situation of the regions of the Community (1984)) point to a continuation of this trend.

Clear as these facts are, their interpretation gives rise to controversy. Whilst some lines of thought following for instance Fourastié's (Le Grand Espoir du 20ème Siècle - Jean Fourastié (1958)) theories might even welcome this phenomenon as a further step in Europe's move towards a post industrial society, other lines of thought looking at trends in the field of international trade tend to interprete this much more as a loss of industrial competitiveness of Europe with regard to both its major industrial competitors, Japan and the USA on the one hand and the New Industrialising Countries on the other hand.

However it may be, a more detailed comparison of employment trends between Europe and its main competitors, particularly the USA, reveals as a major difference that the latter shifts from the industrial to the tertiary sector have been considerably facilitated by high rates of job creation in new branches and products with the result that unemployment rates have remained lower. There is general agreement that new technologies have played a major role in this transition. This leads us to the question - What has been the Community's performance in this field?

New technology

The Commission in its memorandum prepared for the recent Milan meeting of the European Council points out that since 1972 the annual growth rate in real terms of the production of high technology goods in Europe has not exceeded 5 per cent, while the rate in the United States is 7.6 per cent and in Japan 14 per cent. In addition, Europe's declining industrial performance has eroded its trade surplus in high-technology products. Over a 20-year period the export cover of high technology imports into the Community fell from 190 per cent to 110 per cent (1983).

Furthermore in its communication (COM 85/350 final: Towards

a European Technology Community) to the Brussels meeting of the European Council in March 1985 on strengthening the technological base and competitiveness of Community industry (COM 85/84 final), the Commission had re-emphasised the view that, while it is not justified to speak of an overall loss of competitiveness or technological gap, there are worrying trends in several branches with a high technological intensity such as information technology, telecommunications, biotechnology and new materials.

The Community is not willing to accept this deteriorating trend which in the long run will undermine the very economic basis of its international competitiveness and prosperity. It has launched several important initiatives and actions in order to respond to the technological challenges facing it. These include:

- The European Strategic Programme for Research and Development in Information Technologies (ESPRIT) whose objective is to strengthen the base available to manufacturers and users of information technology through encouraging and promoting collaborative research and development in information technologies between industry and research institutions. The estimated cost of the first five years of the programme is 1.5 billion ECU of which the Commission is supporting 50 per cent of the cost.
- Seven multi annual research and development programmes (BRITE) in the fields of energy, biotechnologies and "basic technological research and application of new technologies".
- An action programme in the field of telecommunications. Within this programme six action lines have been defined:
  a) establishment of medium and long term objectives at Community level;
  b) definition and implementation of an R&D programme (RACE);
  c) broadening of the terminals market and development of Community solidarity towards the outside world;
  d) joint development of transnational parts of the future telecommunications infrastructure within the Community;
  e) intensive use of modern telecommunication techniques for the advancement and development of infrastructure in the least favoured regions of the Community;
  f) progressive broadening of those parts of telecommunications equipment markets which are dominated by carrier procurement.
- Within the framework of this telecommunications action programme, the programme for research and development in advanced communications in Europe (RACE) had been recently adopted with the purpose of enabling the transition from a narrow-band Integrated Services Digital Network (ISDN), which already exists in part or is being developed, to the more powerful Integrated Broadband Communications (IBC) by 1995. The Commission believes that promoting and supporting modern European-wide telecommunications infrastructure will make it possible to create a continental market with common standards

capable of supporting the scale of production required to enter
world telecom and technical markets.
- A plan for the transnational development of the organisation of
a framework for innovation and technology transfer which,
interalia, has been instrumental in supporting the
establishment of three associations:
a) The Standing Conference of European Local Authorities
(STCELA);
b) The European Venture Capital Association (EVCA);
c) The European Association for the Transfer of Technologies,
Innovation and Industrial Information (TII).

What Has Been the Impact of Industrial and Technological Change on the Regional Development of the Community?

Before going into details let me state that systematic analysis
of this problem has become an integral part of Community regional
policy. Firstly every two and a half years the Commission
publishes the so-called 'periodic report' on the social and
economic situation of the regions of the Community, in which the
principal features of regional development from a Community point
of view are presented. Secondly being aware of the fact that
other policies may play as important a role as regional policy
itself, the Commission has adopted the principle of carrying out
continuous evaluation of the regional impact of Community
economic and sectoral policies, with a view to either influencing
them or to accompany them financially, be it to strengthen their
positive effects or to compensate for any negative effects
particularly on employment.

Regional impact of industrial change

As regards industrial decline in general, a recent study of the
Centre for Urban and Regional Development Studies in Newcastle-
upon-Tyne (Industrial Decline in the Regions of the European
Community) carried out for the Commission shows that over the
period 1974-1982 this has been widespread because almost three
quarters of the Community's level III regions registered some
decline in industrial employment. Severe decline however, with an
annual rate of more than 2 per cent, affected mainly the
traditional industrial regions in the Community, particularly the
North, the North-West, the West Midlands of England, parts of
Scotland, Wales and Northern Ireland, Lorraine, Nord-Pas-de-
Calais in France, the Saar and parts of the Ruhr in Germany, the
Western part of The Netherlands, and Wallonie.
    The decline of certain industrial branches has had a
particular regional impact:
- In the first place, there is textiles and clothing. This branch
of industry has lost jobs since 1950 and this at an ever

accelerating rate (annual loss between 1950 and 1960 was 0.8 per cent; between 1960 and 1970 it was 1.7 per cent; and between 1970 and 1980 is was 3.4 per cent). The impact is particularly important for regional development in the Community since textiles and clothing is, in terms of jobs, the largest manufacturing subsector and geographically widespread. It is interesting to note that the textile areas in north-west and central Europe have been worst hit, while on the contrary, employment in the peripheral regions (defined in Centrality, Peripherality and EEC Regional Development, a study carried out for the Commission by David Keeble and others, University of Cambridge 1983) has even increased in some cases.

- In second place comes shipbuilding. Here too there was a continuous decline in employment since the 1950s. In contrast to the textiles and clothing sector, the shipbuilding sector is smaller and not so geographically widespread. The worst decline in employment is evident in UK shipyards, followed by France and Germany.

- In third place is the iron and steel industry. Decline in this sector is more recent. Between 1955 and 1974 employment still increased by some 100,000. This was the period when the integrated coastal steelworks developed, importing their raw materials and largely exporting their finished products. Since 1975 however, there has been a major change: overcapacity and falling prices have led to the need for major capacity reductions together with closures and reductions in employment. All Member States' steel industries are affected, although over a long period the UK's losses were especially high here too.

On the whole this analysis would seem to show that peripheral regions have done better than the traditional industrial regions of Europe. Several factors must however be taken into account:

- in some of the least favoured areas (Italy, Greece), official job statistics do not always give an accurate picture of the sectoral employment situation (for example: the existence of special employment maintenance mechanisms, such as Cassa di Integrazione);

- their industries are generally young and supported by government intervention;

- the economies of those regions are indirectly protected by monetary adaptations. The continuous devaluation of the lira and drachma for instance, has without any doubt contributed to maintaining competitiveness in the principal industries both in and outside the Community. The macroeconomic cost of such a policy is however high: it does not contribute towards the convergence of economies in Europe but rather reinforces the danger of 'l'Europe à deux vitesses'.

The Regional Impact of Technological Change

As technological change as such is a broad and complex term, it is difficult to assess its regional impact. The Commission in its framework of regional impact assessment approach has therefore focussed on that aspect which, as I explained, has received major policy attention in the Community, namely new information technology (NIT).

Studies (see mainly the effects of new information technology on the less favoured regions of the Community: Centre for Urban and Regional Development Studies, University of Newcastle-upon-Tyne: Regional Policy Series no. 23) carried out by the Commission show that there is a close correlation between the level of socio-economic development and the level of NIT equipment: higher developed central regions are by far better equipped than less developed peripheral regions. A similar pattern appears as regards the use of NIT-services: the higher density of population and economic activities in the more central regions of the Community favours the use of these advanced services whilst sparsely populated low activity areas discourage intensity of use. The predominance of small and medium-sized firms in those areas is often a further hindering factor.

What are the prospects for the future? All the evidence available to the Commission points to advanced telecommunications developments following demand. This tendency will be further reinforced by the fact that in more and more countries the telecommunications organisations responsible for NIT services are either being privatised or at least have to apply principles of commercial management. So, unless corrective action is taken, infrastructure for the new technologies will tend to follow established demand patterns and concentrate in the most developed regions of the Community in the same way as railways and roads did in the past. This would be all the more regrettable as telecommunications services are, by definition, much more distance independent than were the classical communication means in the past. This fact provides responsible planning authorities with a great chance to improve the traditional geographical pattern of communications infrastructure equipment.

The Role of Community Regional Policy

What has been the contribution of Community regional policy towards overcoming the problems of industrial and technological change?

Policy in the past.

Until the end of last year the ERDF was divided into two sections. Firstly a quota section comprising 95 per cent of the

Fund which was reserved to support national regional policies. This section worked through project financing based on application by Member States. Although some of this finance was used for technology related projects the bulk of it went to classical infrastructure projects. The Commission had little possibility to influence the orientation of this section towards fields of major Community interest.

From this point of view the non-quota section, although it counted for only 5 per cent of the Fund, constituted an important step. Within this section the Commission, with a view to allowing for the regional effects of other Community policies, established on its own initiative specific regional development measures, defining the policy fields to be covered, the areas benefitting and the type of measures to be put into operation. Such measures were launched in order to facilitate the reconversion of areas affected by the decline of steel, shipbuilding and textiles industries. Other measures were devoted to facilitate the adaptation of areas affected by the enlargement of the Community. These measures amounted to 1 billion ECU and covered more or less 20 per cent of the most industrialised areas of the Community affected by serious industrial decline. As the finance available can only be used for the creation of alternative jobs, the non-quota section presented an important contribution to fostering industrial change within the Community.

These specific measures represent also an important contribution to technological adaptation. New types of aids were established with a view to mobilising the indigenous potential of regions, mainly represented by small and medium-sized enterprises and creating a favourable climate for innovation and the introduction of new technology. To this end aids were provided in respect of product and processing innovation, management organisation matters, market studies and better access to risk capital.

Policy in the future

A new era has been opened up by the new ERDF regulation which came into effect on 1 January 1985. The main change consists of much stronger emphasis on the Community interest of Community regional policy in general and Fund action in particular. In order to achieve this, several new mechanisms have been introduced:
- the system of national quotas has been replaced by a system of ranges;
- project financing will be increasingly replaced by programme financing;
- each financial contribution will be assessed on its Community interest according to criteria established in the Regulation.
Within this new framework the place of new technologies will be important.

115

Community programmes related to new technologies

The first possibility for the Community to ensure the benefits of new technologies for less favoured areas consists of establishing 'Community Programmes'. Community programmes are according to the new Regulation the specific instrument to "provide a better link between the Community's objectives for the structural development or conversion of regions and objectives of other Community policies". Like the former non-quota measures they are established at the initiative of the Commission. In its working programme for 1985 the Commission has announced two such programmes which are clearly geared towards promoting technological progress in the least favoured areas:

1. The Community Programme for advanced telecommunications in less-favoured regions. This programme should provide a better link between the Community's regional policy objectives and the objectives of Community Telecommunications Policy which now are clearly defined. It should ensure that less favoured peripheral areas adequately participate in the benefits which advanced telecommunications can provide to economic development. On this basis the community programme envisaged might cover the following fields:
   - establishment of modern telecommunications infrastructure (infrastructure able to link the regions to a transnational backbone system, digitalisation, establishment of data networks and of cellular mobile telecommunications);
   - introduction of advanced services (high speed data transmission, teletex, telefax, videotex, videoconferencing, videophone, satellite business systems and cellular mobile radio);
   - demand stimulation such as awareness campaigns, publicity programmes, user seminars, colloquies and including regional and local telecommunication planning.

As telecommunications corporations' approach is generally demanded and less favoured areas often lack demand, it is clear that demand stimulation will have to play an important role within this programme.

2. The Community programme for energy. This programme is intended to link the objectives of Community regional policy to those of energy policy which have also been clearly defined. It will concern the better exploitation of energy resources with a high local and regional content and more particularly
   - the exploitation of alternative energy sources;
   - the rational and efficient use of energy;
   - demand stimulation including regional and local analysis and planning.

The local impact of operations will be measured on the basis of employment, use of energy and diffusion of new technologies. Although the relation of this programme to high technology is less apparent than for the previous one, it nevertheless exists, because the energy fields covered are those which can assure an important contribution to the local and regional technology base.

National programmes and projects

All national applications for ERDF financing, be it in the form of programmes or projects, are now assessed for their Community interest on the basis of the criteria set out in the new Regulation. The Commission will ensure that among the parameters used for applying the criteria modern technologies receive all importance they deserve.

Indigenous potential

Particular mention should be made of the fact that within the new Regulation the instrument of indigenous potential development, which before was limited to areas covered by the non-quota measures, can now apply to all Fund areas.
This increases considerably the possibilities of small and medium-sized enterprises to benefit from technology transfer and application of new technologies to new products and procedures.

Conclusions

In the light of the above some broad conclusions can be drawn as to the specific contribution of Community regional policy to industrial and technological change.

There is general agreement that Europe's deficiency in new technologies does not lie so much in the field of fundamental research but more in the transformation of the results into industrial processes and marketable products. The Community's regional policy through the instruments described above is one means to facilitate this transformation. This function is all the more important, since Community regional policy does not only cover some dispersed areas but the whole or almost the whole of certain less prosperous Member States (Ireland, Greece and Portugal), or at least important parts of them (Italy and Spain).

Likewise there is general acceptance that the process of innovation depends to a large degree on the stimulation of receptiveness of all active elements, economic, social and political at the local and regional level. Community regional policy through mobilisation of indigenous potential disposes of an appropriate device to create a favourable environment for the introduction of innovative processes at the regional and local level.

Finally regional aid by definition is not bound to specific sectors or branches and therefore does not run the risk of supporting outdated industrial structures. As pointed out above, in its typical Community form (Community Programmes) it cannot be used for internal restructuring of declining industries but has to be put towards the creation of new economic activities.

Community regional policy is therefore a powerful tool for fostering the process of industrial change and technological change in those areas of the Community which need it most.

CHAPTER EIGHT

TECHNOLOGY PARKS AND INTERREGIONAL COMPETITION IN THE FEDERAL
REPUBLIC OF GERMANY

Eike W. Schamp, Professor Department of Geography, University of
Göttingen, West Germany

The Race of Nations

The crisis of the world economy in the 70s led all industrial
nations to reconsider, correct and readjust their industrial
policies to the new conditions of the world market. As trade
liberalisation already accepted in the GATT sessions prevented
states from the extensive use of tariff regulations in trade,
governments shifted their interest towards structural policy
measures. The export-oriented economies like the Federal Republic
of Germany (FRG) felt particularly threatened, on the one hand by
the rise of the newly industrialised countries (NICs) which
caused a considerable competition in mass production branches of
industry like textiles, electronics and steel. By the extension
of the international capital market, however, the multinational
enterprises were enabled to shift the labour-intensive parts of
production to certain developing countries (Fröbel et al. 1977).
In the early 80s, in addition the international debt crisis has
been felt as another threat to German exports. On the other hand,
Japan and the US proved to be serious competitors in modern
products. The nations highly integrated into the world market,
such as the FRG, were thus forced to make a quick adjustment to
their industrial structure, especially in view of the promising
high technologies. A new race of nations had been born (Junne
1984).
    According to their internal economic and social conditions
the industrial countries are choosing different ways to procede.
There was, for example, a considerable rise in the research
expenditures of Japan and the FRG during recent decades. In the
FRG the share of the expenditures of the GDP have risen from 2.1
per cent (1969) to 2.8 per cent (1985). Thus the race of the
large industrial nations turned out to be a research race.
International competition is now mostly understood as a
competition for innovations in products and processes. Despite
the comparatively high R&D expenditures a technological gap
between the FRG and its competitors is suspected (Börnsen et al.
1984). It is assumed that for different reasons - such as the

119

degree of concentration of industry and the negative attitude of the scientific community towards economic issues in the 70s - knowledge has not trickled down at an optimum pace from the places of research to the economic units. This has been the starting point for a new structural policy in the FRG since the end of the 70s.

As far as an economic race that involves pioneers and imitators in the sense of Schumpeter is concerned, we may regard the international spread of political instruments in the light of the innovation diffusion theory. As supposed by Junne (1984, 138) the competitors might feel themselves compelled to take over and make use of the respective new instrument of promotion which is used by the rival neighbour. It must be emphasised here, that the international race is carried out between nations and not between multinational enterprises. The following chapter reports on the use of such a new instrument of promotion in the FRG: the technology or science park.

The International Spread of Technology Parks

The definition of the industrial park has led to an extensive but somewhat unsatisfactory discussion (Barr 1983) which might be taken up again for the technology parks. We should avoid such an attempt at the moment. We are rather going to use the notions 'technology park' and 'science park' synonymously and introduce the following working definition:
The technology park is an industrial park with special characteristics,
- it is designed for high-tech firms only,
- it shall assist the adoption of technological innovations through close contacts with public research institutions,
- it shall serve as a seedbed for new enterprises.
The latter function is characteristic of Technology Parks in the FRG and leads to a time limitation on the presence of enterprises in the park.
Science parks are an American invention and first spread in the United States. The first one was the 'Stanford University Industrial Park' founded on the site of Stanford University in 1948. It offered production facilities to new mobile technology based firms. The idea of the science park is thus closely connected with the concept of the industrial park that has already been successful in the United States and in Great Britain. In 1956 the Research Triangle Park that was concentrated more on research than on production came into existence (OECD 1984, 35). From both experiences we may conclude that technology parks may have different aims. In both cases, however, it took a considerable time to attract a large number of enterprises and to create new jobs. In 1961, for example, there were 25 enterprises with 11,000 employees in the Stanford University Industrial Park and today there are about 17,000 employees. From this fact

follows that first, larger enterprises could settle in the park within a short time and second, as compared to the 400.000 new jobs which have been established in Silicon Valley during the 60s and 70s, this park has little significance as an impetus for regional development (Saxenian 1983).

By the end of 1960 about 50 science parks had been established in the US. Their number has been increased by roughly 25 to 30 during the 70s and today more than 150 are estimated to exist in the US (Henckel 1984, 3). However, more than 50 per cent may be considered as a failure as they have attracted few enterprises. During the beginning of the 70s the idea of science parks spread to Great Britain where two universities established parks of their own: the Cambridge Science Park in 1973 and the Heriot Science Park in Edinburgh in 1974. For a long time the first park did not contribute to what nowadays is called the 'Cambridge Phenomenon'. It took ten years before the park became successful (Segal Quince & Partners 1985, 81). It was not until the beginning of the 80s that new parks were established in Great Britain. Depending on the definition of 'science park' there are varying estimates as to their number. As estimated by Quince (Allesch and Fiedler 1985, 67) some 14 science parks are open at present and another 13 are in various stages of planning or implementation.

Other countries have been more cautious of establishing science parks. In France there is just the 'Parc International d'Activités de Valbonne Sophia Antipolis', founded around 1970; other parks near Lyon are in the stage of planning. Sweden, too, has had some science parks since 1979. Among the large number of industrial parks that have been founded in The Netherlands during the last few years there is just one proper science park: the 'Business and Technology Centre Twente' (Fiedler et al. 1985).

In the FRG technology parks are a very recent phenomenon as the first one was established in Berlin in 1983. But soon, their number was increasing at an enormous speed. The Berlin Centre for Innovation and New Enterprises founded in 1983 was followed by five technology parks in 1984 and another eleven by the middle of 1985 (Wiwo 23/1985, 44). That this is a real boom may be shown by the continuously changing and contradictory information on technology parks: for example in February 1985 five technology parks were reported to be in the stage of implementation in the Federal State of Lower Saxony, and a further seven in the stage of planning. While it was being announced that four other communities had given up their plans one of them reported the opening of its new technology-oriented Centre for New Enterprises in the month of June. Thus, we can only estimate that 65 technologyparks havebeenrealised tillnoworarebeing planned.

The Position of Technology Park-Policy in the FRG

The following points will help to reveal briefly why technology parks as an instrument in economic development have been introduced comparatively late into German politics but nonetheless are now having a boom.

1. Former efforts of the Federal Government to raise the competitive capacity of the economy aimed to improve the general social framework without direct intervention into national market structures. In the 60s, the educational system had been reformed by means of establishing new universities, opening up the 'education reserves' in the rural areas etc., thus increasing the human capital. In the 70s, the growth of public research expenditures aimed at developing new technologies. However, priorities were laid on large-sized projects such as nuclear power, big computer centres and space research (Bruder 1980, 24). Thus the structural policy had an impact on only a restricted number of big companies. It may be that this was one of the reasons why the Federal States started their own technology policies from 1976 onwards (Schütte 1985).

2. As a result of the world economic crisis during the 70s it was recognised both on the regional and local levels that the regional policy practised so far had lost its basis. One should not deny its success in the earlier growth period when regional policy aimed at opening up all economic resources by improving regional infrastructure and fostering new investments, which led in most cases to the establishment of branch plants. At the end of the 70s, however, the interest of regional policy had to shift towards safeguarding the existing small and medium-sized enterprises. The reason for their competitive weakness was identified as their conservative behaviour towards innovations. As an innovation deficit was especially assumed to exist in peripheral regions (which was to be confirmed in a later study by Meyer-Krahmer et al. 1984) a more defensive policy was applied to improve their innovative capacity by subsidising the employment of highly qualified staff and establishing consultancy agencies for innovations and technology.

3. Thus, there are three reasons why the industrial park policy proved to be unsuccessful in the seventies. First, there were no more mobile enterprises to attract and second, policy was now directed more towards the protection of existing enterprises. But third, the offer of plots, buildings and services could not be very attractive because German entrepreneurs aim to be independent and purchase a plot of their own. Furthermore, local authorities offer industrial zones well equipped with infrastructure practically for nothing. Thus, while in the United States most new plants are reported to be located in industrial parks (Barr 1983, 423), these parks did not work efficiently in the FRG.

4. It was not until the beginning of the 80s that the rapidly

growing unemployment enforced the reconsideration of all policies oriented towards existing big or small enterprises. As shown in Table 1, the FRG has been a late-comer in high unemployment rates compared to other industrial nations.

Table 1. Unemployment in the FRG 1980 - 1985

| Year | unemployed (in million) | rate of unemployment | | | | |
	FRG	FRG	UK	F	US	Japan
1980	0.89	3.3	6.3	6.4	7.1	2.0
1981	1.27	4.8	9.2	7.8	7.6	2.2
1982	1.83	6.9	10.6	8.8	9.7	2.4
1983	2.26	8.4	11.5	8.9	9.6	2.7
1984	2.27	8.4	12.0	10.1	7.5	2.7
1985*	2.21	8.9	13.4	9.6	6.8	2.4

* Wiwo 39/85, 4 (month of July and August, different base of calculation)
Source: Stat. Bundesamt 1985: Statistik des Auslandes, wichtigste westliche Industriestaaten 1985

According to American experience mainly the smaller enterprises were said to contribute to the reduction of unemployment (Birch 1984). Consequently, hope was set on newly established enterprises. This seems to be paradoxical because it is constantly being stressed that the establishment of new enterprises in the FRG is hampered by the attitude of the German employees and by the socio-political environment. But along with the rapid increase in unemployment there was a boom of new establishments that surpassed the number of insolvencies at the end of the last decade. Unfortunately, this process has not been sufficiently reported so far (Schatz 1984). One of the main reasons seems to be necessity: it is interesting to note that during the early industrialisation of Germany unemployment and underemployment also gave an essential impetus to the founding of new establishments (Kocka 1975, 55).

Consequently, political measures are nowadays designed to favour the setting-up of enterprises. For the first time in the economic policy of the FRG, programmes have been set up to foster, first, new firms in any branch whatsoever, for example by strengthening the capital formation of new entrepreneurs. Since July 1983, the formation of technology based enterprises has been promoted in certain pilot areas (Fig. 1). Public authorities of all territorial levels stimulate the formation of venture capital companies. From 1983 to the end of 1984 the number of such companies rose from one to 24, founded by private banks, public saving banks, large-scale industrial corporations (i.e. Siemens)

Fig. 1. Regional Distribution of Newly Founded Enterprises* in 1984 (* promoted by ERP-funds, from Institut der deutschen Wirtschaft)

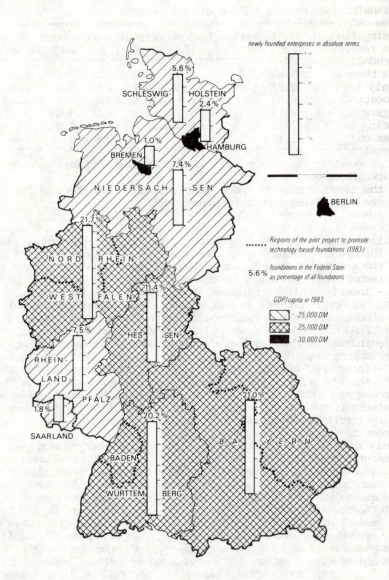

and insurance companies (Allesch and Fiedler 1985). Second, other programmes are intended to stimulate the transfer of scientific knowledge from technical and traditional universities and colleges to the enterprises. A programme to subsidise the labour costs of highly qualified R&D personnel in small and medium-sized firms has been working since 1979 to facilitate technology transfer through the employment of people from universities. On the other hand, since the end of 1984 authorities promoted the delegation of young academically trained persons from the firms to research labs for at most three years. Those programmes are predominantly directed at the small and medium-sized enterprises. As they aim at improving the general economic conditions they only incidentally influence locational conditions. To improve locational factors, a widespread system of technology consultancy agencies has been established and today, technology parks are being built up. Thus, it seems obvious that the technology park has only a complementary function in the control of regional and local industrial change.

Nevertheless, there are two functions assigned to technology parks in Germany nowadays:
- the first one is directed towards general economic, structural goals: new high-tech enterprises are expected to secure the future capacity of German industry in international competition. Nobody, however, can tell what a high-tech industry is really going to be in future times (Breheny et al. 1983, 62).
- the second function is dedicated to the regional policy: the technology parks shall help to abolish
  - interregional disparities between the federal states in the 'wealthy' south and the 'poor' north. As shown in Fig. 1, however, new enterprises are concentrated in the south of the FRG.
  - intraregional disparities between the areas of growth and depression and
  - intermodal disparities between the major metropolitan areas in the FRG.
Consequently, technology parks are used as a means in the interregional competition of locations in different ways.

If Silicon Valley is continually brought into discussion in order to justify the establishment of technology parks it is not only to legitimate the new policy. It also makes obvious the expectations of success connected with this kind of policy:
1. New technology-oriented enterprises, so called 'spin-offs', may arise from universities and colleges, thus enforcing a quick technology transfer into the economy.
2. These enterprises should belong to high-tech industries, i.e. make use of a new basic innovation and thus contribute to structural economic change.
3. These enterprises are small in the beginning but probably grow quickly, i.e. they can offer new jobs.

While it is generally concealed that regions like Silicon Valley have prospered predominantly by considerable military orders and that for the most part the plants have been built without any public locational policy, it is nowadays believed by

Table 2. Scale of joint locations from the industrial park to the science park

industrial park	centre for new enterprises	technology park	science park
(1)	(2)	(3)	(4)
availability of plots to settle permanently mobile and new plants in the FRG often producing for the lokal market; a minimum of services is offered	availability of plots and buildings for rent to settle new enterprises and give them a higher chance of survival by services offered	additional to (2): the neighbourhood of research labs is expected to attract high-tech firms and to facilitate high-tech spin-offs; production is possible but firms should leave after a certain period of time	additional to (2): the neighbourhood and contacts to R&D facilities should promote spin-offs and give new high-tech firms the chance to develop a new product/process to maturity without producing it at the science park site

Source: Taylor 1983, Krist 1984

the political authorities of all territorial levels that these conditions of a 'seedbed' for young high-tech firms can be created artificially by Technology Parks. As a means of a supply-side economic policy technology parks in the FRG are characterised by offering
- a favourable location, i.e. by subsidised rents on space, close contacts to research labs and other high-tech enterprises,
- different forms of financial aid, i.e. venture capital,
- a hothouse-environment for newly founded enterprises, i.e. by taking over business duties (accountancy, market analysis, marketing) and providing infrastructure which can be used in common (typing office, reception service, telecommunication).
It is striking that these are mostly the same means that have been designed to make industrial parks attractive to enterprises. As mentioned above, the basic difference of the technology park concept in the FRG lies in the admission of small high-tech firms, of new establishments for a limited period of time after which the enterprises are supposed to leave the seedbed, and in the postulated contact with science and research. Thus it is not astonishing that in the FRG there is a wide range of differently

designed parks from a simple industrial park for new enterprises with a more verbal pretention towards high-tech firms to the pure science park without any production facilities at all.

The Spread of Technology Parks in the FRG

For two reasons it is difficult to give a detailed sketch of the spread of technology parks in Germany: first, nearly every day we hear of new plans to build up a technology park, second, it is difficult to identify real technology parks because the planning concepts leave open a number of questions so that we can hardly differentiate between a real technology park and a simple centre for new establishments.

By the end of 1984 altogether six centres for high-tech firms and new establishments could be counted. Up to the middle of 1985 another ten centres were opened (Table 3, Fig. 2). As Table 3 shows, most parks are financed or co-financed by local authorities, i.e. parks are used as an instrument for local economic promotion. Therefore technology parks are in general initiated by local government officers or local politicians. They found private agencies to manage the park. When the authorities are in a hurry and have limited funds they often make use of empty industrial buildings or schools. However, the co-financing by the federal state, the local authorities, local banks, and chambers of commerce means that all take interest in the technology park policy. Consequently the objective of the technology park is not too precisely determined and thus not limited. Either it is oriented to the basic innovation of microelectronics or completely left open as long as promising technologies - whatever these may be - are developed by the enterprises. It is an exception if a technology park is oriented to the research labs or the economic structure of the respective region. This, for example, is the case in Heidelberg where enterprises engaged in biotechnology and biotechniques have been established. However, beside Cologne, Munich and Berlin, Heidelberg is the fourth centre for genetic engineering research in Germany where there is a collaboration between public large-scale research laboratories and the chemical industry (in this case: BASF) that is financially supported by the Federal Ministry for Research and Technology (Wiwo 26/85, 86). Thus, the technology park for young and innovative enterprises adds to a broader policy to build up a national research centre on gene-technology. As may be expected, even during the starting phase of the park, big companies are already beginning to take over some of these innovative firms. We should realise that in this way technology parks only subsidise big corporations (Eisbach 1985).

Other examples of appropriate parks may be found in Dortmund and Stuttgart, both cities in densely industrialised regions with a high proportion of metal processing, mechanical engineering or car industry. In close cooperation with technical universities

Table 3. Technology Parks in the FRG (June 1985)

Location	year of establishment	type of agency	branches	number of existing enterprises	planned	spin-offs from university	presence limited in time	sum of investments (million DM)
Berlin	1983	1	different	26	50	x	x	5.5
Aachen	1984	1	different	13	35	x	x	3.0
Bochum	1984	3	different	5	9	x	-	.
Karlsruhe	1984	1	microelectronics	20	25	x	x	15.5
München	1984	1	microelectronics	1	.	.	.	.
Schwerte	1984	3	different	5	.	-	-	.
Bonn	1985	3	different	45	200	-	-	20.0
Dortmund	1985	1	material flow systems and material techniques	15	34	-	x	10.0
Hannover	1985	1	microelectronics	9	35	x	x	.
Heidelberg	1985	1	gene-, biotechnology	9	25	x	x	10.7
Hildesheim	1985	2	different	9	25	-	-	12.0
Köln	1985*	1	different	2	.	.	.	6.0
Mannheim	1985	1	microelectronics	1	.	-	.	.
Saarbrücken	1985	1	different	6	.	-	.	30.0
St. Georgen	1985	3	production engineering	5	10	-	-	.
Stuttgart	1985*	4	production engineering	8	20	-	-	13.0
Syke	1985	4	microelectronics	13	13	-	-	3.3

type of agency: 1. joint venture by municipality, banks, chambers of commerce
    2. municipality only
    3. private owner
    4. banks and Federal State

* = building-up phase
x = yes
- = no
. = no answer

Sources: Fiedler et al. 1985, Wiwo 23/1985

and other institutions of technical research, the enterprises of the parks are preoccupied with, for example, automatic steering devices and production technique in general.

Thus, the spread of technology parks involves quite different phenomena. In contrast to the park concept in the US or UK especially, the parks situated in large cities serve as a 'seedbed' for new enterprises. This means that enterprises have to leave the seedbed after three years or, as is more common, after five years. In some technology parks they are not allowed to produce but are restricted to development carried out in cooperation with research labs. This kind of park comes close to the science park in a narrower sense, like the one in Heidelberg and Hannover. Anyway, the authorities hope that though research and production activities of an enterprise are separated, the producing enterprises are kept in the neighbourhood, thus strengthening the regional economic structure. This should be the case, too, if enterprises leave the park.

In other cases there is no time limit on residence. Here and there it is expected that the closing down of firms or their move caused by rapid growth and lack of space gives way for new establishments (e.g. Hildesheim). This concept is generally followed in locations which are not well equipped with universities or big research laboratories. With rare exceptions, only technology parks in metropolitan areas are situated in the vicinity of research centres. Nevertheless, it is expected by the management of other parks that distance does not have any significance for establishing close contacts between enterprises and the knowledge centres (Fig. 3).

Assumptions on the Regional Impact and the Use for Publicity

If the establishment of technology parks is meant to ensure an improved competitive capacity of locations on the regional, national and international level we may expect some detailed ideas of their mode of action. However, a short and incomplete examination of the arguments raises some doubts about the effectiveness of this instrument.

- First, it is assumed that new high-tech firms will pioneer structural change. But we are not at all sure about which technologies are really promising. Are the public managers of a technology park expected to know what is going on in the future? The resulting ideas about high-tech industries are different in detail though they follow a general line, i.e. one chooses industries which are sufficiently known, particularly all fields of microelectronics. This may be exemplified by two cases: In the first and rare case all enterprises belong to the same field of technology. This is true for Heidelberg where eight enterprises work in gene-, medical- and biotechnologies and just one on computer software. Five of them are the rare

Figure 3. Technology Parks in the FRG (June 1985)

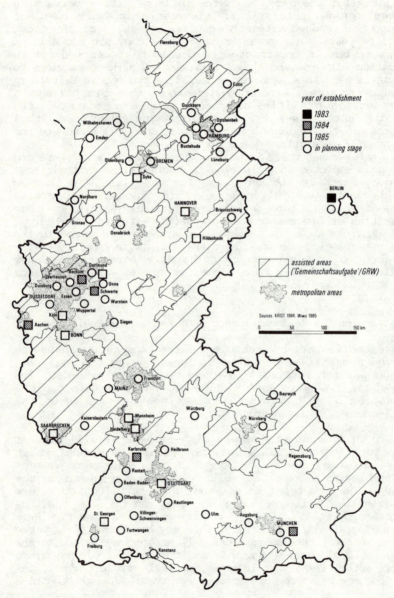

year of establishment
■ 1983
▨ 1984
□ 1985
○ in planning stage

BERLIN

assisted areas
('Gemeinschaftsaufgabe'/GRW)

metropolitan areas

Sources: KRIST 1984, Wiwo 1985

0   50   100   150 km

but desired spin-offs of the university and research centres. Other parks with a high proportion of spin-offs are those of Aachen, Berlin and Karlsruhe (Wilhelm et al. 1985). In the second and more frequent case the microelectronic industry is represented by software firms, i.e. business services in its proper sense. This is the case with Syke near Bremen where eight enterprises out of 13 offer computer software, engineering services and consultancy. There is apparently no spin-off. Apart from the fact that only a small number of computer software enterprises can be established in technology parks it is this very branch which is expected to exhibit an extensive destructive competition followed by concentration in the near future.

- It is continually being pointed out that the chances of survival of new enterprises are considerably raised by the 'seedbed'-function of the technology park and that this alone contributes to the desired structural change (Keune and Nathusius 1977, 32). We know that the death rate of small firms within the first five years amounts up to 90 per cent. We know nothing, though, about a diminished death rate in German technology parks because, first, they are still too young for this kind of experience and second, only firms which can be expected to survive pass the strict selection for a park. It is this kind of enterprise with a high propensity to survive, then, which is subsidised by the technology park.

- It is assumed that there is a large amount of knowledge in universities and research labs which can be disseminated predominantly by means of spin-offs. As it seems necessary that spin-offs keep close contacts with these knowledge centres during the development phase of products till they are marketable, they are said to be sensitive to distance. There are indeed examples of a distance decay of knowledge transfer: in 1981 nearly 10 per mille of the producing enterprises in the metropolitan area of Karlsruhe were provided with patents from the Centre for Nuclear Research. In the remaining parts of the Federal State of Baden-Württemberg the licence rate was only 1.8 against 1.3 per mille in the rest of the Federal Republic (Wiwo 18/84, 74). But apart from such anecdotal information there is no systematic analysis of the extent and distance of spin-offs. As shown in a new study (Schamp and Spengler 1985) the research contacts of traditional universities with the economy differ between sciences. Chemistry is an example with far-reaching contacts, reaching all metropolitan areas in the FRG. We may conclude from this that a spin-off is not dependent on spatial proximity in a state which is densely populated and highly equipped with infrastructure.

- Although in the FRG 16 technology parks are going to be in operation by the end of 1985 and another 50 are in the building

131

phase or in the stage of planning they cannot be expected to have a direct impact on the desired structural change, neither on the national nor on the local level. As shown in Table 2 there are in each park between 10 and 20 enterprises with 4 to 10 employees in the average. Furthermore, the space in the technology parks is restricted to 1,000 to 5,000 square metres. That means that the direct effect on employment is negligible. Big plants of companies in some traditional industries such as the car industry or the electrical industry still create more new jobs than any technology park. The parks do not even reach the necessary 'critical mass' for a take-off in regional development that has been estimated to be 15 new establishments each year for a longer period (Shapero after Keune and Nathusius 1977, 36). The effect is even weaker if we consider, for example, the fact that only 42 per cent of the enterprises in the technology parks of Lower Saxony are really new establishments (though 97 per cent are young ones, Schröder 1985, 113). Additionally, recent estimates of the annual amount of new high-tech enterprises only come to 250 (Eisbach 1985), at the maximum to 500 new firms (Krist 1984) in the FRG. This may explain why already today there is so much competition between municipalities for young high-tech firms.

- Thus the hope for multiplier effects can only lie in the distant future. But still a number of municipalities expect early results from their policies thus failing to see that their famous examples like the Stanford University Industrial Park or the Cambridge Science Park had to accept a considerable time lag between the start and effectiveness.

- Finally, the technology park fails as a means of the new innovation-oriented regional policy. As shown in Fig. 2 and Table 4 there is neither a reduction of the south-north differences in the FRG nor a compensation between the metropolitan and the depressed areas. This is not surprising as most authors stress the following locational factors for a successful technology park: an attractive location in which to live, nearby universities and research laboratories which are renowned for appropriate natural and technical sciences, resources of high-skilled labour, and a future-oriented psychological environment, that means: growth breeds further growth.

Thus, technology parks do not seem to be an appropriate instrument of a rational regional policy. On the contrary, we may assert that the establishment of a technology park is a sign of being up-to-date for territorial governments which act as crisis managers with varying strategies and success. The regional policy practised so far created an almost ubiquitous infrastructure, e.g. for transport, education, sports. Today, local authorities are looking for a further offer to attract new entrepreneurs.

Table 4. Technology parks in the FRG according to types of region

type of region	in operation	technology parks under construc- tion or in the stage of planning	total number
non-assisted areas	13	37	50
thereof			
metropolitan areas	12	24	36
assisted areas ('GRW')	3	11	15
thereof			
metropolitan areas	1	1	3

Source: Kirst 1984, Wiwo 23/85, 40

Albach was right to point out that the dynamic entrepreneur according to Schumpeter needs a counterpart: the dynamic politician (1979, 347). Indeed, most technology parks arose from the activities of politicians and have been created as a form of supply-side economic policy. Only this may explain on the one hand, why so many technology parks have been established in the Federal State of Baden-Württemberg that has the lowest rate of unemployment and the highest growth rate while neighbouring 'poorer' states like Rheinland-Pfalz venture almost nothing. It may also explain, on the other hand, why remote communities like Syke in Lower Saxony are among the first to open and fill up a technology park. Here, the dynamic politicians cooperate on different territorial levels, as shown by the joint ventures. While the individual communities would have to provide considerable funds the share of financing by the regional and Federal Governments means only a small burden for them. We may estimate the total costs of technology parks created so far at up to 100 million Marks (DM). However, programmes which do not affect locations raise much more funds. For example, the Federal Programme for the Founding of technology-oriented Establishments has made available 325 million DM for the period from 1983 to 1985, and the general Federal Programme for the Founding of Enterprises provided 1.4 thousand million DM in 1983 alone (Wiwo 3/85, 55).

Under these circumstances technology parks are probably expected to have little direct impact on regional development. As a psychological means some of the parks may be successful especially by attracting bigger enterprises to the municipality. Thus they are rather a sign which is supposed to stress the progressive image of a community or Federal State. Unfortunately, such sorts of sign loose their effectiveness if there are too many of them.

References

Albach, H. (1979) 'Zur Wiederentdeckung des Unternehmers in der wirtschaftspolitischen Diskussion' In: Zeitschrift Ges. Staatswissenschaften, 135, 533-552

Allesch, J. and H. Fiedler (Eds.) (1985) Management of science parks and innovation centres. Berlin

Barr, B.M. (1983) 'Industrial Parks as Locational Environments: A Research Challenge' In: F.E.I. Hamilton and G.J.R. Linge (Eds.) Regional Economies and Industrial Systems: Spatial Analysis, Industry and the Industrial Environment, Vol. 3, Chichester, 423-440

Birch, D. (1984) 'The Contribution of Small Enterprise to Growth and Employment' In: H. Giersch (ed.) New Opportunities for Entrepreneurship. Symposium 1983, Tübingen, 1-17

Börnsen, O., H.H. Glismann and E.J. Horn (1984) 'Fällt die Wirtschaft der Bundesrepublik Deutschland technologisch zurück?' In: Die Weltwirtschaft, Heft 2, 1984, 171-183

Breheny, M., P. Cheshire and R. Langridge (1983) 'The anatomy of job creation? Industrial change in Britain's M4 corridor' In: Built Environment, 9, 61-71

Bruder, W. (1980) 'Innovationsorientierte Regionalpolitik und räumliche Entwicklungspotentiale - zur Raumbedeutsamkeit der Forschungs- und Technologiepolitik des Bundes' In: W. Bruder and Th. Ellwein (Eds.) Raumordnung und staatliche Steuerungsfähigkeit, Politische Vierteljahresschrift, 20, Sonderheft 10/1979, 235-253

Eisbach, J. (1985) Gründer- und Technologiezentren - Sackgassen kommunaler Wirtschaftsförderung. Bremen, Progress-Institut für Wirtschaftsforschung, Studien 1

Fiedler, H., K. Fromm, M. Krüger and Ch. Scheffen (1985) International Directory of Science Parks and Innovation Centres, Draft, Berlin

Fröbel, F., J. Heinrichs and O. Kreye (1977) Die neue internationale Arbeitsteilung. Reinbek

Henckel, D. (1984) Science parks, Gründerzentren. Berlin, Difu Kurzinformationen

Junne, G. (1984) 'Der strukturpolitische Wettlauf zwischen den kapitalistischen Industrieländern' In: Politische Vierteljahresschrift, 25, 134-155

Keune, E.J. and K. Nathusius (1977) Technologische Innovation durch Unternehmensgründungen. Eine Literaturanalyse zum Route 128 Phänomen. Bifoa-Forschungsbericht, 77/4, Köln

Kocka, J. (1975) Unternehmer in der deutschen Industrialisierung. Göttingen

Krist, H. (1984) Gründer- und Technologiezentren als Instrumente zur Verbesserung der regionalen Innovations- und Anpassungsfähigkeit. Fraunhofer Institut für Systemtechnik und Innovationsforschung, Referat im Rahmen des 'Seminars für Planungswesen' der TU, Braunschweig am 13. Dezember 1984

Meyer-Krahmer, F., R. Dittschar-Bischoff, U. Gundrum and U. Kuntze (1984) Erfassung regionaler Innovationsdefizite. Bonn, Schriftenreihe 06 'Raumordnung' des Bundesminister für Raumordnung, Bauwesen und Städtebau, Nr. 06.054

OECD (1984) Industry and University. New Forms of Co-operation and Communication. Paris

Saxenian, A. (1983) 'The Genesis of Silicon Valley' In: Built Environment, 9, 717

Schamp, E.W. and U. Spengler (1985) 'Universitäten als regionale Innovationszentren' In: Zeitschrift für Wirtschaftsgeographie, 29, 166-178

Schatz, Kl.W. (1984) Die Bedeutung kleiner und mittlerer Unternehmen im Strukturwandel. Kieler Diskussionsbeiträge 103, Kiel

Schröder, K. (1985) 'Technologieparks' in Niedersachsen eine Analyse der Ziele, Konzepte und Implementierung. Unpublished diploma dissertation, Göttingen

Schütte, G. (1985) 'Regionale Technologieförderung in der Bundesrepublik Deutschland' In: Zeitschrift für Wirtschaftsgeographie, 29, 145-165

Segal Quince & Partners (1985) The Cambridge Phenomenon. The Growth of High Technology Industry in a University Town, Cambridge

Taylor, T. (1983) 'High-technology industry and the development of science parks' In: Built Environment, 9, 72-78

Wilhelm, H., H. Corsten and P. Peckendraht (1985) 'Erste Analyse ausgewählter Technologieparks in der Bundesrepublik Deutschland. Ergebnisse einer Befragung' In: Zfbf Schmalenbachs Zeitschrift für betriebswirtschaftliche Forschung, 7/8, 618-630

SZ = Süddeutsche Zeitung
Wiwo = Wirtschaftswoche

CHAPTER NINE

PHILIPS IN THE WORLD
A VIEW OF A MULTINATIONAL ON RESOURCE ALLOCATION

Dr. J. Muntendam, Corporate Director Philips International B.V.,
Eindhoven, The Netherlands

Introduction

Philips has been in the marketplace of the whole world for nearly
95 years. Philips was compelled to look beyond the national
borders very early in the history of the company. The small Dutch
homeland, even so many years ago, could not provide the big
market, which was needed for the early lamp products.
Multinationalism therefore, became a characteristic for Philips.
The company now has its own 'national organisations' in more than
sixty countries, which control up to 420 factories. The total
number of employees is 346,000. In terms of sales (60 billion
Dutch guilders in 1985) Philips ranks the fifth place on the
world list of electronics companies.

Table 1. Employees per geographical area. Situation 31 December
1985

Area	Employees (1,000)
Netherlands	71
Europe (excl. Netherlands)	151
USA and Canada	55
Latin America	30
Africa	5
Asia	28
Australia and New Zealand	6
Total	346

Source: Philips Annual Report 1985

A Corporate Research Organisation, with eight laboratories in
Western Europe and the United States, is the driving force behind
the rapid technological development and product innovation within
Philips. That characterises Philips as a technology-based

industry. The availability of advanced technology and a strong innovation process are essential preconditions for the continuity and expansion of the company.

In fundamental research (taking some 1.25 per cent of turnover) Philips has approximately 4,000 employees. Half of these research personnel are employed at Eindhoven. The other research laboratories are located in Redhill (UK), Limeil-Brèvannes (France), Hamburg and Aachen (West-Germany), Brussels (Belgium), Briarcliff Manor and Sunnyvale (USA). Development activities (some 6 per cent of annual turnover) are located also for 50 per cent in The Netherlands.

The far reaching diversification of the product range is a third fundamental characteristic of the company. Philips at present has nine Product Divisions: (1) Lighting; (2) Consumer Electronics (audio and video); (3) Major Domestic Appliances (refrigerators, washing machines); (4) Small Domestic Appliances (shavers, coffee grinders etc.); (5) Medical Systems (X-ray, MRI etc.); (6) Industrial and Electro-acoustical Equipment (test and measuring, numerical control, electronic microscopes); (7) Defence Systems; (8) Telecommunication and Data Systems (computers, banking systems, word processors); (9) Components Division (tubes, ICs). This contrasts with other big electronic companies, which confine their activities to one or only a few related products or product groups. Concerning Philips, over the years the emphasis has shifted to horizontal diversification, which means supplying the fullest possible range of related goods and services for a world market. To realise this, it is frequently necessary to embark on co-operative activities and joint ventures with other companies, to obtain supplementary know how. The huge investments for Research and Development is another reason or necessity to co-operate with competitors in order to acquire basic knowledge or realise standardisation.

Stages

The history of a nearly century-old company like Philips is, of course, characterised by ups and downs. Nobody could expect that a future multinational was created when Mr. Gerard Philips and his father Mr. Frederik Philips, the banker from Zaltbommel, located in the middle of The Netherlands between the rivers Rhine and Waal, bought an old factory in Eindhoven in 1891 to start the manufacturing of incandescent lamps. The availability of this building was the main reason to choose Eindhoven as the location for their new-born company. Later on this location turned out to be an attractive one, because of the availability of cheap and abundant labour. Until 1930 Philips grew rapidly. The company was expanding in lamps, and related products such as radio tubes, X-ray equipment, which were all made in Eindhoven. Due to World War I Philips was forced to produce their own glass and other components. Even then exports from The Netherlands had already

Figure 1. History of Philips

1891	Eindhoven
Till 1930	Expansion products/Vertical integration
1930-1940	Protectionism --- European industrial activity
1945-1957	Protectionism and labour-shortage
1957-1973	EEC
	Booming economy / labour-shortage
1973-today	Oil crisis
	Far East
	Restructuring

assumed such proportions that Philips set up its own sales companies in twenty-four countries, six of them outside Europe. Initially these were serviced by the factories in Eindhoven. Measures taken in the twenties and the thirties to limit international trade led to more and more production outside The Netherlands.

At first, this production consisted of the local assembly of components imported from Eindhoven. Gradually, however, these components were produced abroad too in growing quantities. In important (European) countries the selling company, together with the factories established there, started a local product policy. In that period important acquisitions took place in various European countries, which increased the product capacity rapidly. However, these plants abroad produced exclusively to meet local demand.

The years immediately following the Second World War were characterised by still stronger restrictions in the field of international trade than those of the thirties. It became desirable for Philips to achieve a geographically decentralised production system. Economic activity exploded in that time in most countries of Western Europe. Therefore activities were started in a growing number of countries. One of those countries was Austria where there was plenty of labour. Looking at the industrial map of Philips, Austria is even now still over-represented in relation to sales.

Within The Netherlands the same development took place. Due to a severe labour shortage near Eindhoven Philips started to recruit, as many other companies did, employees from the Mediterranean Base. But only after new production facilities were started outside Eindhoven. Gradually the number of plants elsewhere in The Netherlands grew. In 1972 there were already 90 Philips factories in The Netherlands. The situation now is given in Fig. 2.

A specific specimen of this shift was the establishment of a shaver factory in Drachten, located in the province of Friesland, one of the traditional problem areas in The Netherlands. This plant was for that reason welcomed and subsidised by the national government, as were some other newly established Philips plants in the northern provinces. The small town of Drachten, however,

Figure 2.

PHILIPS FACTORY LOCATIONS

NETHERLANDS

o   situation 1985
•   closed/deconsol. after 1972

had never had any industrial activity and Philips started a factory right from scratch. In a rather short period, with the help of Philips Eindhoven, they had succeeded in a huge industrial operation. Today a factory with almost 2,000 employees with all kinds of specialisations makes a full range of shavers.

The next important phase for Philips was the formation of economic blocks in the sixties and the seventies. At that time most of the factories were just producing on the basis of local for local. With the EEC gradually becoming more forceful, it became necessary to use the economies of scale. That had an enormous impact on the number and location of the factories. For many countries this meant that they were forced to coordinate their own production not only within the home market, but also with the export business. Moreover, development and product policy could no longer be left to the various countries, but called for an international approach. Gradually the factories were integrated in International Production Centres as the backbone of the various Product Divisions. The geographical structure of the company changed over those years in a more product-oriented one.

Up to 1973 Philips had a major expansion, but then the oil crisis and later on competition from the Far East disturbed the picture, and Philips was confronted with stagnation. A restructuring and reallocation of the resources was necessary, not only in Europe but also in the rest of the world. Due to rapid technological developments, based on micro electronics, streamlining of the organisation became a continuous process. Ultimately, Philips is aiming for a sufficient market position in the three large market areas, Europe, the United States and the Far East, in order to continue its role on the leading edge of development. This strategy demands various levels of production. To achieve the necessary flexibility, Philips is considering that large central production units should be surrounded by satellite factories. The development of both product and production process would take place in such a central manufacturing unit, the centre of know how and innovation. For example, the fluorescent lamp-factory in Roosendaal (The Netherlands) is already the main centre for supplying production technology to various satellite factories (pure production-units) in other parts of the world.

Market-Share

Concerning the allocation of production facilities it is difficult to indicate general trends, because of the strong diversification of the product range. One cannot compare the manufacture of TV receivers, videorecorders, test and measuring instruments and professional MRI installations. Nevertheless a few basic lines can be drawn.

It is increasingly considered important to have a sound base in the United States. More than 40 per cent of world demand for electronics is located there. Moreover, various centres of competence for new professional technological developments are developed there. This market is eminently suitable for the introduction of new electronic products. It is obvious that Philips should have substantial production facilities in the United States, particularly in Silicon Valley, where the continuous process of innovation in the chip-manufacturing industry makes one's presence essential. For that reason Philips bought Signetics Inc. in 1976. As a result the number of employees, employed in the United States, grew rapidly. In 1972 it was 4 per cent of the total number all over the world; in 1985 it had grown to 15 per cent. In the same period the U.S. sales

Figure 3.

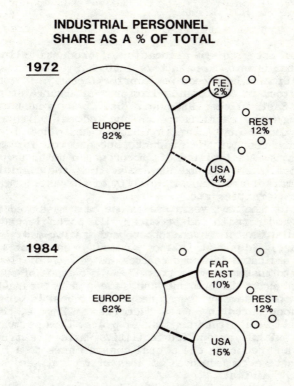

**INDUSTRIAL PERSONNEL
SHARE AS A % OF TOTAL**

1972

EUROPE 82%

F.E. 2%

REST 12%

USA 4%

1984

EUROPE 62%

FAR EAST 10%

REST 12%

USA 15%

increased to 29 per cent of the total net sales (Fig. 3).

In spite of the dynamism of the innovative power of American companies in high-technology sectors, the Americans have had to yield ground to the Japanese consumer electronics. The market share of American companies for ICs has declined to the benefit of Japanese companies. As a general trend the production of a large number of industrial (especially consumer-)goods has shifted over to the South East Asian Newly Industrialised Countries (NICs). Countries such as Korea and Taiwan have grown spectacularly. This development along with the strong position of Japan and the recent advance made by China, seems to be resulting in a shift of the world's economic centre from the Atlantic Basin to the Pacific Basin. For an international company like Philips, this implies it should have direct contacts with these centres. Philips cannot neglect or withdraw from these countries, as the Far East has a wide range of interesting production capabilities: availability of a dedicated workforce, great flexibility and a good business climate. The number of employees, employed by

Philips in the Far East, was only 2 per cent of the total number in 1972. In 1984 it had grown to 10 per cent.

The Future

Decisions relating to the allocation of production have to be taken with the utmost care. Factors influencing the decision are changing in time. Labour cost was a prime factor in 1891 to start the lamp factory in Eindhoven and to produce a range of products in the Far East in the 1960s and 70s. Today wages are less important, as their influence on the (added) value of most products is decreasing. Automation often takes place not because of the labour cost of the product, but due to, for example, quality reasons. Availability of labour was a major argument in the allocation in the 50s and 60s, today it is the flexibility in the labour market and the availability of infrastructure (good suppliers, universities etc.).

Trends for the next years are on one hand mass production of components and certain end-products like portable radios and pocket calculators but on the other hand 'tailor-made' software (and hardware). To put it in another way: economy of scale is pointing to dedicated low cost centres and economy of flexibility is pointing to customising centres. Those two types of production centres can exist in one and the same product range. Tuner production and assembly of printed-circuit boards for TV sets will be concentrated in a few factories in Europe, the final assembly will be close to the market place. This permits us to make use of existing production facilities while streamlining the production and reducing the social consequences to a minimum. Also the negative consequences of loss of capital and know-how should be kept minimal.

This, however, refers to re-allocation of production, a process we have been going through over the past five years and will continue to go through for the next few years, the result of many years of allocation of production on local for local basis or because of labour shortage. For newer products like the compact disc player the production remains for many years concentrated in one factory (Hasselt in Belgium) until sufficient economy of scale is reached. The same goes for the components like motors, ICs, lenses etc. produced in factories all over Europe. The sum of economy of scale, infrastructure (suppliers), government incentives, cost of forwarding, flexibility and wage levels will prove if Europe is still competitive on the production side.

But there is more to it. Product- and process-innovation is of paramount importance to remain competitive. The research and development provides the necessary innovation. The technology push, combined with market pull should result in the right product at the right time. An innovation centre requires, however, the right environment, the neighbourhood of universities, suppliers of hardware like high precision machinery, optical industries but also IC design centres and components. One cannot live without the other. The fading out of consumer electronics in the USA means the fading out of the component industry for consumer electronics. The lack of scale in

Figure 4.

SALES
Per area as a % of total

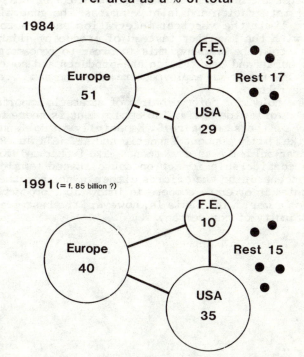

1984

Europe 51

F.E. 3

USA 29

Rest 17

1991 (= f. 85 billion ?)

Europe 40

F.E. 10

USA 35

Rest 15

professional electronics in Europe compared to USA and Japan means a serious drawback for the component industry in Europe. For a company like Philips this leads to the conclusion that we should be at the centre of gravity of professional electronics: the USA. With 29 per cent of worldwide sales in the USA and only 15 per cent of the industrial activity, it is clear that the latter one should increase especially in the ever more important sector of professional electronics. On the other hand in the consumer sector the presence in the Far East will be reinforced as that region has become the centre of gravity of consumer electronics. Meanwhile sales of only 3 per cent of world sales should pick up also in view of the ever increasing income per capita in the region in order to match more or less the 10 per cent industrial activity Philips had in 1984. Europe's industrial share of 62 per cent will further decrease in the future but will remain above the share in sales, the latter gradually going down from 51 per cent in 1984 to 40 per cent in the 1990s (Fig. 4).

It will be a smaller part, however, of a far bigger pie, estimated at about 90 billion guilders in 1991, when Philips will celebrate their centennial. What that means, in number of employees, will not only result from technological change, but also from other factors like social/employment conditions, degree of vertical integration and joint ventures. The general trend today is to limit the vertical integration and create a co-makership with the supplier instead of producing all kind of components within the company, this in order to concentrate all resources, human and capital, on end-products and some basic components. The result in employment remains the same, be it on different payrolls.

The same holds for joint ventures of strategic importance to compete on a worldwide basis. The employment is there but not necessarily on the same payroll. Especially in Holland joint ventures like Philips Dupont Magnetic and APT (ATT and Philips Telecommunication) but also divestments like Duphar and NKF steel have had a considerable impact on consolidated (employment) figures. Concentrating the efforts either or not together with other companies in order to compete in innovation, production and marketing on a worldwide basis is, however, the best guarantee for the continuity of the company.

CHAPTER TEN

INNOVATION DECISION-MAKING IN SMALL AND MEDIUM-SIZED FIRMS
A BEHAVIOURAL APPROACH CONCERNING FIRMS IN THE DUTCH URBAN SYSTEM

Drs. J.A.A.M. Kok and Dr. P.H. Pellenbarg, lecturer and senior
lecturer, Department of Geography, University of Groningen, The
Netherlands.

Introduction

One of the focal points of interest in economic geography is the
structure and change in industrial systems. Attention on
industrial change is mainly on new firms, technology related
questions and innovation. The central theme within this chapter
is innovation decision-making, which is only one aspect of
industrial change. Innovation decisions concern a fundamental
renewal of the firm's production, its organisation or its
marketing strategy. Such decisions can stretch the firm's life-
cycle and keep the industrial system up to date.

Recently there has been a large amount of research
concerning industrial innovation (Molle 1985). In economic
geography attention was paid to the spatial distribution of
innovative firms in several countries (Oakey, Thwaites and Nash
1980, Ellwein and Bruder 1982, Kok, Offerman and Pellenbarg
1984a). In these investigations differences between central and
peripheral regions were discovered in innovativeness in small and
medium-sized enterprises. Many questions were raised about the
relationship between these spatial disparities and the character
of the innovation process.

In The Netherlands the spatial aspects of the innovation
theme were mainly investigated by means of the structural
approach. This means that at an aggregate level the
investigations are focussed on the characteristics of a region
that might stimulate innovative behaviour (N.E.I. 1984, Alders
and De Ruyter 1984). Although there are similarities with the
actual distribution of innovative firms in The Netherlands, no
attention has been paid in these investigations to the
relationship of the characteristics of an area and the actual
innovative behaviour of firms in these areas. There was not any
special attention given either to the urban areas, where the
innovation processes should tend to cluster, according to most of
the well-known innovation diffusion theories.

In this chapter we will deal with the impact of local
environmental conditions on the innovative activities of the

individual entrepreneurs in some selected urban areas in The Netherlands. This line of research presumes a behavioural approach which concentrates on the innovation decision making activities of the entrepreneurs, especially with regard to the proces of obtaining and using relevant information. The main problem within such research is to link this information process to spatially determined characteristics. On the other hand we investigated the locational and mobility patterns of the innovative firms and their possible consequences for the Dutch urban system.

Before presenting the results of our empirical investigation we will outline shortly the operational definition of industrial innovation we used throughout our research and comment on the figures on industrial innovation in three agglomerations (Rotterdam, Utrecht, Arnhem/Nijmegen). These figures will be compared with the 'average' Dutch innovation figures. A section with conclusions and forthcoming lines of research will conclude this paper.

An Operational Definition of Industrial Innovation

Like the above mentioned investigations of Oakey et al. (1980) and Ellwein and Bruder (1982) we will, for the greater part of this article, confine ourselves to innovation in small and medium-sized firms (up to 100 employees). This restriction has to be understood from the argument that these small and medium-sized firms (SMEs) constitute a sector of the economic system where quick reactions to changes in demand and technology are possible, and at the same time the dependence on local environmental conditions is rather great. In other words, SMEs are of a special importance in the process of industrial change, and their activities can possibly be manipulated by government action with regard to the conditions of their location. After this, we will refer to these conditions with the term 'production milieu'.

Our engagement with SMEs means that basic innovations (for instance: the jet engine, synthetic fibres, transistors, chips) fall outside the scope of this study. In modern western society those basic innovations are the outcome of planned research and development, conducted by big multiplant firms, sometimes operating jointly with government institutions. SMEs seldom, if ever, participate in this process. They usually engage only in 'normal' innovations, which can be derived from basic innovations.

This leaves us with the question what a 'normal' innovation really is. The existing literature about all aspects of innovation does not provide clear answers to this seemingly simple question. Since Schumpeter introduced the innovation concept, many different definitions have been used, most of them rather vague. Schumpeter himself defined innovation as "doing things differently in the realm of economic life", which may serve as an example of this vagueness (Schumpeter 1939). We adopted the more direct definition that is currently used in the official publications of the Dutch government: "Innovation is the development and successful introduction of new and improved products, services, production and distribution processes" (Innovatienota 1979).

Because the definition had to be used as a means of identification and classification of innovations, it needed further operationalisation. In the first place this concerned the types of innovation, where we distinguished product innovations (services also understood as products), innovations of the production process, organisation innovations, and market innovations. Secondly, we distinguished between four levels of innovation, to prevent mixing up really important renovations of a production system with minor adjustments to new trends. These four levels can be described as follows:
- basic innovation: something completely new which has a general economic impact (as stated, this falls outside the scope of our study);
- primary innovation: innovation produced by a firm which can be based on a basic innovation and is as such new to the world;
- secondary innovation: innovation produced by a firm which is not new to the world, but only for the country where the firm is located;
- tertiary innovation: an adaptation or improvement.

The four types and four levels of innovation combine to sixteen different ways of appearance of innovation within firms. Fig. 1 illustrates this idea. Some of the combinations, especially of the market-type, need further explanation. Primary market innovations are directly related to primary product innovations; we will count such developments only once, as product innovations. Secondary market innovations occur rather frequently in our calculations; they are synonymous to initiating export activities. Tertiary market innovations concern geographical expansions within the home market. It is necessary to stress that tertiary innovations as we defined them do not correspond to what is generally referred to by the word innovation. We nevertheless think it is useful to distinguish this innovation level, because it is a rather important one when small and medium-sized firms are dealt with. To prevent confusion, figures in this article will be given separately for the three innovation levels.

The Characteristics of Innovative Firms in The Netherlands. A Comparison of a National and an Urban Sample

On the basis of Fig. 1 we have investigated the innovativeness of the three already mentioned urban areas. In this section these results will be compared with the results that emanated from our research with regard to the regional distribution of innovative firms in The Netherlands as a whole (Kok, Offerman and Pellenbarg 1984b).

The figures for the urban areas are based on three different samples: a sample of 3.3 per cent for Rotterdam, 7.7 per cent for Utrecht and 14.3 per cent for Arnhem/Nijmegen. The areas are shown in Fig. 2. The different percentages reflected our differing expectations as to the number of innovative firms to be found. The samples included all independent firms in the

Figure 1: Types and levels of innovation.

Innovation		Basic		Primary		Secondary		Tertiary
Invention	Economic Feasibility	Product Process	Diffusion	Product Process	Diffusion	Product Process	Adaptation	Product Process
Idea		Organisation Market		Organisation Market		Organisation Market		Organisation Market
Res.	Development			Exchange of Information				
		BIG FIRMS						
				Small- and Medium- Sized Firms				

ISIC-sectors of manufacturingindustry, wholesale, transport and business services with more than 1 and less than 100 employees.

The sample that was taken for The Netherlands as a whole had the same characteristics. In this case it was a 1 per cent sample. In both cases the innovative characteristics of the firms were established by means of a short telephone inquiry. The samples and the main results are shortly summarised in Table 1.

Table 1. Comparison of the national and urban samples.

	Urban areas	Netherlands
1. Total number of firms	600	607
2. Useful for further research	472	461
3. Innovative firms	175 (37%)	147 (32%)
4. Innovations	242	179
5. Pr./sec. innovations	68 (28%)	60 (34%)
6. Innovation rate (4:3)	1.4	1.2

Figure 2: Investigated regions.

In Table 2 all the recorded innovations are shown in an 'innovation matrix' which gives a breakdown for all recorded innovations to types and levels. To make these results compatible the innovations are shown as a percentage of the total number of innovations in the urban areas and The Netherlands.

Table 2. Comparison innovation matrix for urban areas (UA) and The Netherlands (N); number of innovations in % of total innovations.

| | Primary | | Secondary | | Tertiary | | Total | |
	N	UA	N	UA	N	UA	N	UA
Product	6	3	13	6	13	14	32	24
Process	1	–	4	1	18	15	23	16
Organisation	–	–	3	1	25	35	28	36
Market	–	–	7	16	9	7	17	24

From the Tables 1 and 2 it appears that the general innovation rate is slightly higher in the urban areas. However, concerning the level of innovativeness we just conclude that the urban areas have less primary and secondary innovations. Concerning the types of innovations the share of product and process innovations is lower in the urban areas than in The Netherlands as a whole. On the other hand the urban areas have a greater share of organisational and market innovations.

Looking at the characteristics of the firms we have to note that there are no striking differences between the innovating firms in the urban and national samples. Most of the innovations are accounted for by manufacturing industries, particularly the chemical, metal and electronic industries. Small firms produce less innovations than big firms and young firms less than old ones. On the other hand it appeared in the urban sample that employment growth based on innovations is primarily accounted for by the younger and smaller firms.

Some interesting conclusions can be drawn from the comparison of the national and urban samples . First the general innovation rate in the urban agglomerations is slightly higher than the national average. At the same time we have to admit that the innovation level was higher in the national sample. Nationally, 34 per cent of the innovations were of the primary and secondary kind (60 out of 179), whereas in the urban agglomerations only 28 per cent of the innovations were primary or secondary in nature (68 out of 242). The difference is determined to a great extent by the many tertiary organisation innovations in the urban agglomerations. More important is the higher rate of secondary market innovations (translation: starting export activities) in the urban agglomerations, compared to the national figures. This phenomenon can perhaps be explained by the different mix of industrial sectors in both samples: the urban sample accounts for 40 per cent wholesaling firms and 37 per cent manufacturing industry firms, against 31 per cent and 44 per cent respectively in the national sample.

A cautious conclusion from the figures we presented could be that the innovation rates in the urban agglomerations do not clearly exceed the national average. A slightly higher number of innovations is more or less offset by a lower level of innovation. This conclusion however, needs some adjustment when the individual urban agglomerations are concerned. The percentage of innovating firms in Utrecht is markedly higher than in Rotterdam and Arnhem/Nijmegen.

These conclusions support our earlier conclusions about the regional distribution of innovative firms in The Netherlands (Kok, Offerman and Pellenbarg 1984b). Differences in innovativeness for the central parts of the country and the periphery do exist, but we already noted that innovation occurs in big and medium-sized cities as well as in the smaller towns around them (Fig. 3). Although the agglomerations stand out with a large absolute number of innovative firms, the concentration is

Figure 3. Distribution of innovative firms by COROP-areas.

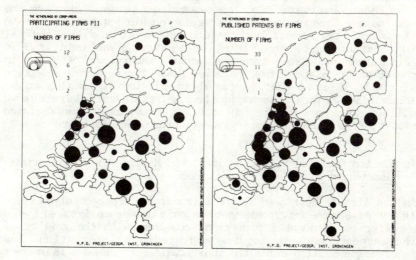

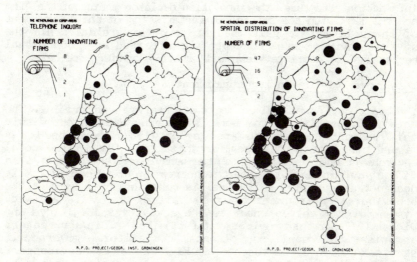

by no means impressive and even less so when by relative standards. On these grounds, the distribution of innovative firms was typified as a rather deconcentrated phenomenon, not exclusively tied to the big or medium-sized cities.

These observations raise several questions. What are the reasons for this obviously loose relationship between innovative firms and the urban environment and what are the spatial implications in terms of locational patterns and mobility? To gain insight in these questions we have interviewed 90 firms in the urban areas (30 firms in each). These firms were a selection of the 175 firms that appeared in our urban samples to be innovative. Before discussing the results of the interviews we will present some theoretical notions about innovation decision making, from which the interview-structure were derived.

Innovation Decision Making in Small and Medium-Sized Enterprises. The Innovative Firm in its Environment

As was already mentioned in the introduction the central question in our research-programme was the role of the environment as a supplier of relevant information, combined with the ability of the entrepreneurs to acquire and use this information. The information available for the individual entrepreneurs is present in certain structures represented by other firms, organisations and persons in the environment of the firm under study. The key-question in our research is the relationship between this information structure, the innovation decision making of the firm and the link with the spatial characteristics of the environment of these firms. To gain insight into this relationship the several environmental concepts which served as a conceptual background for our interviews will shortly be discussed here.

The Innovative Firm in its Environment

The basic environmental concepts are represented in Fig. 4. The total or objective environment is the broadest concept within this respect. McDermott and Taylor (1982) describe it as "everything outside the focal organization". Within this total environment one can distinguish the behavioural environment. This is the environment of which the entrepreneur is conscious. It does not necessarily influence his behaviour. The part of the behavioural environment which is part of the daily activity system of the entrepreneur is the activity space (Lloyd and Dicken 1977) or task environment (Dill 1958). This environment comprises the suppliers, clients, competitors, government, services etc. As is represented in Fig. 4 this task-environment is split up into an information space and an action space. Also the firm under study can be seen as part of the environmental structure, as the decision-making unit. In that case it is labelled the decision space.

Figure 4: Basic environmental concepts.

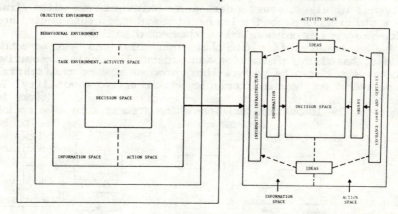

Figure 5: The task environment of the innovative firm.

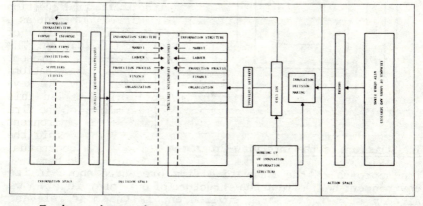

To investigate the connection between innovative decision-making and the characteristics of the environment of a firm the task environment is the most important conceptual level. In our preliminary research the most important impulses for innovation-oriented behaviour in a market-oriented economy turned out to be the exchange of information and goods combined with the individual characteristics of the firm and entrepreneur. These three are, respectively, represented by the information space, the action space and the decision space. With regard to the decision space it has to be noted that most SME's are characterised by single decision centres, in most cases the entrepreneur himself.

In Fig. 5 we take a closer look at the task environment of the innovative firm. It must be kept in mind that the elements that fill the information and action space must give in fact an answer with regard to the motives of innovation and the support during the innovation process. Information and orders, as external impulses, start a decision-making process within the firm which will probably lead to innovations. This decision-making process is schematically represented in Fig. 6.

Fig. 6 makes it clear that innovation can have several causes, based on information and orders. The decision-making process itself can be routine-like, process-like or cyclical and is dependent on the characteristics of the entrepreneur.

Figure 6: The innovation decision-making process.

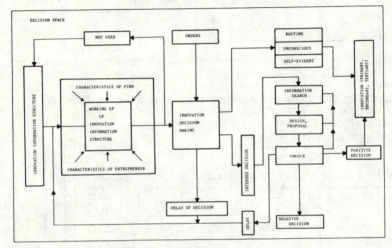

The Structure of the Interview-Programme

The approach of the innovative firm presented above in its environment has defined the structure of our interviews. As we already stated, the task environment is the most relevant conceptual level. On this level we find the exchange of information and goods which forms in most cases the basis for innovative behaviour. In the questionnaires which were used for the interviews these crucial interaction processes were uncovered by a series of questions concerning the production and information milieus as they were experienced by the respondents. The decision space, the third element of the task environment (Fig. 4) was 'mapped' through questions concerning the most important firm characteristics (nature, size and growth of production, personnel figures etc.) and qualities of the entrepreneur (age, education, business experience, regional origins, etc.). Finally, the major innovation characteristics of the firm were established.

The 90 interviews with innovative firms in the three urban areas under study produced a host of information, not all of which can be presented in this chapter (see also Kok, Offerman

and Pellenbarg 1985). We will concentrate on the following central questions:
- What was the ultimate reason to innovate?
- What sources of information have been used during the innovation-process and what is their spatial 'range'?
- What is the valuation of several aspects of the 'production milieu' for innovating behaviour?
- What are the main characteristics of the successful innovative entrepreneur?

In the next section the answers to these questions, as they were produced by the 90 interviews, will be presented.

Reasons to Innovate

From Table 3 it appears that the most important incentive towards innovative behaviour of small and medium-sized firms is the demand of clients for better or other products or services. This incentive has also the highest priority. It is followed by a need to expand the market share and the diminishing of production costs. This pattern showed no significant difference for the distinguished urban areas.

Table 3: Incentives to innovate.

	Total	Priority		
		1	2	3
Demand of clients, suppliers	68	35	15	18
Expansion market share	54	21	25	8
Diminishing production costs	43	19	18	6
Technological developments within firm	28	4	8	16
Product diversification	27	7	8	12
Changes in the organisation	14	3	7	4
By accident	9	1	1	7
Other reasons	1	0	1	0

These findings stress our earlier conclusion that the most important motive to innovate in SMEs is the relationship between the firm and its clients.

The Search for Information

It has already been stated that information is a crucial aspect in the innovation process. Ellwein and Bruder (1982) in their research in Western Germany already noted that innovation problems mostly turned out to be information problems. We have made a distinction between need for information and sources of information.

The need for information

Corresponding to the market-oriented character of the innovations
it is quite obvious that the subjects on which the entrepreneurs
feel a need to acquire information have to do with the market and
the marketing strategy.

Table 4: Subjects of information need.

Subjects	No. of times mentioned	Priority 1	2	3
Market	62	34	17	11
Marketing	41	7	21	13
Product/service	33	13	11	9
Raw materials	24	11	6	7
Organisational techniques	24	5	4	15
Financial subjects	23	3	5	15
Distribution	20	7	7	6
Labour market	6	2	2	2
Real estate	3	0	0	3
Other	5	0	2	3

From Table 4 it appears that information about the
product/service is less mentioned but has a high priority. This
again stresses the firm-client relationship as an important basis
for innovative behaviour. The position of the financial subjects
is surprisingly low. Finance is often seen as an important
barrier with regard to innovation and had a higher priority in
our national sample. An explanation could be the fact that the
interviewed firms turned out to be the stronger firms in terms of
employment growth, turn-over growth and financial position.

Sources of information

Having discussed the most important needs for information this
section deals with the sources of information used by the
innovative entrepreneurs. Table 5 gives an extensive list of the
various sorts of external consultancy which is used by the
entrepreneurs. It clearly shows some clusters of information. The
first cluster comprises the clients, suppliers, professional
journals and firms. This could be characterised as the informal
information sources or as Oakey stated "the private consultancy"
(Oakey 1981, 43).
        The second cluster has a less informal character. It
comprises the accountants, the banks, Chambers of Commerce and

the external consultancy. The third cluster are the official information suppliers, mainly governmental organisations. This cluster has a very low score. The position of the professional journals is very striking in this list. They are very often mentioned but have a low priority. It seems to be a source that is used as an additional information supplier. At this moment we do not yet know what exactly is the impact of these journals on innovative behaviour.

Table 5: Sources of information.

| Sources | Number of times | Priority | | | | |
		1	2	3	4	5
Clients	69	37	13	8	7	4
Professional journals	55	4	7	10	16	18
Trade-fairs	52	8	14	13	6	11
Informal contacts	48	5	14	6	13	10
Suppliers of raw materials	43	13	17	7	6	0
Accountants	27	5	6	4	6	6
Suppliers of means of production	27	6	4	4	6	7
Internal consultancy	26	6	3	6	6	5
Banks	21	3	3	8	3	4
External consultants	18	3	2	7	2	4
Chamber of Commerce	16	0	2	6	3	5
University (-related)	10	0	2	3	4	1
Books	8	0	0	1	3	4
Ministry of Economic Affairs	8	0	1	3	2	2
Advisory firms	7	0	0	3	1	3
Lawyers	2	0	0	0	1	1
Regional Development Agency	2	0	0	0	1	1
Others	11	0	2	1	4	4

Table 6 shows the spatial scale of the main sources of information specified for the different urban areas. The predominance of the international scale especially with regard to clients and trade-fairs is clear. The overall impression is that the information field for these entrepreneurs is national or international. At the regional level we find some more contacts with the clients, and the informal contacts. It also appeared from this investigation that the external consultancy took primarily place on the regional level. Apparently distance is not an important constraint for the information supply of these innovative firms.

There seems to be no such thing as a regional information system in which these firms are participating for the sake of their innovation decision-making. With regard to the information aspect this leads us to doubt the importance of the urban agglomerations in The Netherlands as pre-eminant information spaces.

Appreciation of the Locational Environment

In this section we will discuss the importance of various aspects of the locational environment ('production milieu') for the innovative activities of the entrepreneurs. The elements under consideration were the labour market, the market area, infrastructure, presence of other firms, social and cultural facilities and government. In the questionnaire there was first a question concerning the importance of each aspect, followed by a question about the appreciation of the same aspect with regard to innovation. Table 7 gives an overview of the importance and appreciation of the various aspects of the production milieu.

Table 6: Spatial scale of information contacts.

Source	Urban area	Regional	Spatial scale National	International	Total
Clients	Rotterdam	6	8	8	22
	Utrecht	0	7	17	24
	Arnh./Nijm.	5	8	10	23
	total	11	23	35	69
Supplier	Rotterdam	4	6	6	16
raw	Utrecht	2	2	6	10
materials	Arnh./Nijm.	0	9	8	17
	total	6	17	20	43
Trade	Rotterdam	1	2	16	19
fairs	Utrecht	0	4	13	17
	Arnh./Nijm.	0	2	14	16
	total	1	8	43	52
Informal	Rotterdam	4	6	7	17
contact	Utrecht	1	6	8	15
	Arnh./Nijm.	3	9	4	16
	total	8	21	19	48

It is obvious that the qualitative and quantitative aspects of the labour market are of great importance for the innovative activities of the entrepreneurs. Other aspects of the locational environment, i.e. the quality of the infrastructural facilities, socio-cultural aspects and contacts with other firms and institutions are important but not necessarily linked with innovations. Although the contacts with other firms (clients or suppliers) are an important incentive to innovation (see Table 3), the presence of other firms in the urban areas is valued as being of low importance. It stresses again the fact that the contact-patterns in most cases extend beyond the boundaries of the urban agglomerations.

Table 7: Valuation and appreciation of aspects of the locational environment.

	Importance			Appreciation		
	A/N	U	R	A/N	U	R
Quality labour market	+	+	+	+	+	+
Quantity labour market	+	+	+	+	+	
Quality infrastructure	++	+		++	+	++
Contacts government	+				+	
Other firms, same sector	+			+		
Other firms, other sector	+					
Soc. cultural facilities	+			+	+	+
Other institutions	++			++	+	+

Note: + 50% of the respondents answered important/good
++ 75% of the respondents answered important/good

A/N Arnhem/Nijmegen
U Utrecht
R Rotterdam

The Innovative Entrepreneur

The last point to be of importance for innovative behaviour turned out to be the characteristics of the innovative entrepreneur. It has often been stated that good entrepreneurship is not dependent upon sectoral characteristics or environmental conditions but primarily on the qualities of the entrepreneur himself. For that reason we tried to create a profile of the innovative entrepreneur. This profile was based on several questions concerning age, education, former jobs, geographical origin and attitude towards entrepreneurship. The profile is based on significant differences for these characteristics with regard to primary/secondary and tertiary innovations.

The innovative entrepreneur is mostly middle-aged (30-50

years) and has been in charge of the company for less than 10 years or more than 25 years. Most of the time this entrepreneur has been in charge of two other firms in the same line of business. Most of the interviewed entrepreneurs were born and bred within the province where the firm is located. This is clearly in contrast to the innovative firms in the northern - rural - part of The Netherlands where most of the managers were raised outside the three Northern provinces (Pellenbarg and Kok 1985). Innovative entrepreneurs unfold many activities outside the company within entrepreneurial circles. They closely follow the market developments on which they react with product and marketing incentives. To them their entrepreneurship is a challenge and last but not least this is possible 'at any time and everywhere'.

Locational and Mobility Aspects of the Innovative Firms

In the preceding sections of this chapter we saw that there was barely a relationship between the locational characteristics of the urban areas and the innovative activities of the firms. An important question at this stage is to what extent this influences the locational behaviour, and more specifically, the mobility pattern of these firms.

Table 8 shows the nature of the location of the firms in the Dutch urban areas. There is difference in the table between the high-level (primary and secondary) innovations and the low-level innovative firms.

Of the total number of 90 innovative firms there were 50 located in the central urban areas and 40 in the suburban municipalities. Almost 50 per cent of the innovative firms were located in industrial areas.

There are some remarkable differences when the level of innovation is taken into account. Of the high-level innovative firms only 17 per cent is located in the inner-city area and the surrounding 19th century districts. For the low-level innovation firms this is 39 per cent. On the contrary the industrial areas contain a higher proportion of high-level innovative firms. This means that the high-level innovative firms spread towards the fringes of the agglomerations. The three distinguished urban areas showed no marked differences with regard to this tendency. The key-question is now in how far this is reflected in the (potential) mobility of these innovative firms.

The moves of the investigated firms are shown in Table 9. Again they have been specified by period and innovation level. With regard to the period 1975-1984, it is striking that 46 per cent of these firms (41 out of 90) have changed their location. This is high compared for instance with figures from recent research concerning all firms and their moves in the northern part of the Randstad area (De Jong and Lambooy 1986).

What is also clear from this table is the fact that

Table 8: Nature of the location of innovative firms.

Location	Total		Firms with prim./sec. innovations		Firms with tertiary innovations	
Central urban area	N	%	N	%	N	%
Inner city	16	18	8	14	8	26
19th century districts	6	7	2	3	4	13
Residential area	5	6	4	7	1	3
Industrial area	23	26	17	24	6	19
Sub total	50	56	31	53	19	61
Suburban municipality						
Central area	20	22	12	20	8	26
Industrial area	20	22	16	27	4	13
Sub total	40	44	25	47	12	39
Total	90	100	54	100	31	100

especially the high-level firms turn out to be the most mobile
firms until 1980. After 1980 the low-level firms turn out to be
much more mobile. These moves can be combined with product-
innovations.

Table 9: Mobility of firms (number of firms moved).

Period of move	Primary/Secondary innovative firms		Tertiary innovative firms	
	N	%	N	%
Before 1970	9	15	3	10
1970-1974	10	17	2	7
1975-1979	16	27	6	19
1980-1984	10	17	9	29
No moves	14	24	11	36
Total	59	100	31	100

Concerning the direction of these moves we have to make clear
that we only counted moves within the agglomerations. The moves
within the agglomerations are predominant. Moves that trespass
onto the central city boundaries are relatively more often
accounted for by the high-level innovative firms.

The reasons for these moves are those which we are familiar with from earlier research. Growth of the firm, inadequate premises and urban renewal aspects are mentioned most frequently. The innovation itself is never mentioned but this is obvious because it is not the innovation itself which causes a move but the impact it has on the character, organisation and size of the production-process. By asking the entrepreneurs about their future migration plans, it turned out that more than 20 per cent had problems with their present location. All of these firms planned to change their location in the future. An important aspect is that almost all of these firms innovated on a high level. Whether or not these plans are definite, a rate of 20-25 percent of firms that have a tendency to move must be designated for the Dutch situation as a very high rate.

Concluding Remarks

In this chapter we have treated several aspects of innovation decision-making in small and medium-sized firms in some selected agglomerations in The Netherlands. The most important conclusions will be summarised in this section.

The innovativeness of firms in the agglomerations turned out not to be higher than in The Netherlands as a whole. The rate of innovative firms was slightly higher but on the other hand the innovation level was lower. For the small and medium-sized firms the most important reasons to innovate were the demand of their clients and market developments. Connected with this market oriented or demand-pull innovative behaviour the information need centred upon the market and the marketing. The most important sources of information had an informal character. This cluster comprised clients, suppliers, trade-fairs and professional journals.

The key-question in this research was to link innovative behaviour with spatially determined characteristics of the local environment of a firm. The main conclusions must be that distance is no constraint with regard to the information-search of the innovative entrepreneur, nor does the agglomeration function as an information or activity space for these firms. A clear example is the fact that on the one hand the contacts with other firms, especially seen as clients, are designed as the most important incentive to innovate and first source of information. On the other hand however these other firms have, according to the respondents, no meaning as elements of their locational environment.

Other aspects of the locational environment, except the quantity and quality of the labour market, have no influence on innovative behaviour. Spatial incentives towards innovative behaviour of SMEs are not tied to well-defined bounded action spaces, for example urban agglomerations, but are connected with the specific spatial characteristics of organisations, persons,

other firms and institutions on which these incentives are based.
In recent years many people have doubted the stimulation of the urban areas on the dynamics of economic life. Ayodeji and Vonk (1980) for instance rejected the validity of the incubator hypothesis for the Dutch urban system. Wever (1984) showed that concerning the birth and death of new firms the 'survival rates' were highest in the smaller municipalities. For three reasons we have to doubt the postulated expectation that the big urban agglomerations are the principal seed-bed of industrial innovation:
1. innovative firms are not overrepresented in the big urban agglomerations;
2. innovative firms located in the big urban agglomerations don't see clear and tight relationships between their innovative activities and characteristics of the locational, urban environment;
3. the innovative firms show already for more than ten years a high migration rate and this tendency will be continued in the near future.
Combined with other research which shows a high attractiveness and economic performance of the so-called Halfway zone or Radiation zone this might include a shift of the economic gravity centre away from the Randstad area. This might leave the urban centres with great problems in the near future, especially with regard to their economic base.

Note

[1] Thanks are due to Dr. G. Ashworth for checking our use of English.

References

Ayodeji, O. and F. Vonk (1980) Metropolitan employment developments, Planologisch Studiecentrum T.N.O., Delft
Alders, B.C.M. and P.A. de Ruyter (1984) De ruimtelijke spreiding van kansrijke economische activiteiten in Nederland, Studiecentrum voor Beleid en Technologie & Planologisch Studiecentrum T.N.O., Apeldoorn/Delft
Boeckhout, I.J. and W.T.M. Molle (1983) Technologische verandering en regionale ontwikkeling, N.E.I., Rotterdam
Dill, R. (1958) 'Environment as an influence on managerial autonomy', Administrative Science Quarterly, 2, 409-443
Ellwein, T. and W. Bruder (1982) Innovationsorientierte Regionalpolitik, Opladen
Innovatienota (1979) Staatsuitgeverij, 's-Gravenhage

Jong, M.D. and J.G. Lambooy (1986) 'Urban dynamics and the new firm; the position of Amsterdam in the Northern Rimcity'. In: D. Keeble and E. Wever (Eds) New Firms and Regional Development in Europe, Croom Helm London, 203-224.

Kok, J.A.A.M., G.J.D. Offerman and P.H. Pellenbarg (1984a) Innovatieve bedrijven in Nederland, Sociaal-Geografische Reeks no. 32, G.I.-R.U.G., Groningen

Kok, J.A.A.M., G.J.D. Offerman and P.H. Pellenbarg (1984b), 'The regional distribution of innovative firms in the Netherlands'. In: M. de Smidt and E. Wever (Eds.) A profile of Dutch Economic Geography, Van Gorcum, Assen, 124-149

Kok, J.A.A.M., G.J.D. Offerman and P.H. Pellenbarg (1985) Innovatieve bedrijven in het grootstedelijk milieu, Sociaal-Geografische Reeks no. 34, G.I.-R.U.G., Groningen

Lloyd, P.E. and P. Dicken (1977) Location in Space, Harper & Row, London

McDermott, P. and M. Taylor (1982) Industrial organisation and location, Cambridge Geographical Studies 16, Cambridge University Press

Oakey, R.P. (1981) High technology industry and industrial location, Gower

Oakey, R.P., A.T. Thwaites and P.A. Nash (1980) 'The regional distribution of innovative manufacturing establishments in Britain', Regional Studies, 14, 3, 235-253

Molle, W.T.M. (Ed.) (1985) Innovatie en regio, Staatsuitgeverij, 's-Gravenhage

Pellenbarg, P.H. and J.A.A.M. Kok (1985) 'Small and medium-sized innovative firms in The Netherlands urban and rural regions', Tijdschrift voor Economische en Sociale Geografie (TESG), 4, 242-252

Schumpeter, J.A. (1939) Business cycles, 1, McGraw Hill, New York

Wever, E. (1984) Nieuwe bedrijven in Nederland, Van Gorcum, Assen

CHAPTER ELEVEN

THE SPATIAL PATTERN OF HIGH-GROWTH ACTIVITIES IN THE NETHERLANDS

Dr. E. Wever, Professor, Department of Geography and Planning,
Catholic University, Nijmegen, The Netherlands

Introduction

Our future depends on our ability to create new technology and
render it commercially viable. This sounds like forcing an
already open door. Nevertheless, economic theory recently had
hardly considered questions like 'how is new technology created?'
and 'what factors constrain or stimulate the process of
technological development?'. Technological development was seen
as an exogenous factor. However, this situation has changed
recently. Economic growth stagnated and our traditional
instruments proved to be inadequate in stopping this stagnation.
In this situation the long Kondratiev-wave, was rediscovered.

In the upswing period of Kondratiev-waves it appears that
'bunches' of innovations enter commercial production on a large
scale. Mensch (1977) and Kleinknecht (1984) indicate that these
innovations stem from the preceding depression. Freeman (1984)
however, relates these innovations to the upswing period itself,
when profit expectations are favourable for taking the risks of
doing something 'new'. At this time however, we do not wish to
enter this debate. In this chapter we prefer to focuss our
attention on the apparent relationship between innovations and
economic growth. 'Bunches' of innovations mostly concentrated in
certain economic sectors, contribute significantly to the
creation of a new upswing period.

Because of the presumed relationship between economic growth
and technology, all Western-European countries nowadays are
seeking to develop new key economic activities. These activities
will, to a large extent, determine their position in
international competition. Countries achieving technological
leadership in the new 'key-activities' can gain strong
competitive positions. Geographically this situation is also
relevant because when an upswing period is based on specific
sectors there is always a possibility that the spatial pattern of
economic activities will change fundamentally. Firstly, because
leaders in the previous period will not necessarily be the
leaders in the next period. Secondly, because the locational

165

conditions of the new key-activities may differ widely from the conditions for the old ones.

An illustration of the first argument is given by Schumpeter (1939). He considered the Industrial Revolution as the first Kondratiev-cycle. Great Britain was the technological leader. The next cycle he placed in the second half of the nineteenth century (steam, railways, Bessemer-steel). The leadership of Great Britain was then already seriously challenged by Germany and the US. The third cycle, in the first half of the present century, was oriented to electricity, chemical and car industries. During this cycle the US took over technological leadership. The next cycle cannot yet be distinguished clearly. Japan however, now seems to be assuming a strong leadership role (Table 1), if we consider the international trade in high-tech products as an indicator of the current fourth Kondratiev-cycle. A shift from an Atlantic to a Pacific economic scenario may be one of the outcomes!

Table 1. Intra-product specialisation* in the production of 'high-tech' products

	1963	1967	1975	1980
USA	.57	.43	.38	.28
EC	.19	.16	.12	.06
Japan	.22	.44	.54	.57

* = (export-import)/(export+import). A country only exporting high-tech products and not importing at all has a specialisation index of + 1.0. A country not exporting and importing them all has an index of - 1.0

Source: Wemelsfelder 1985, 222

Technological development can also have a different impact on individual regions. It is not necessarily the same kind of regions that will attract the 'key-activities' in the different Kondratiev-waves. An illustration is given by Aydalot (1984). In his opinion, the older industrial regions, which specialised in activities identified with the third wave, are not attractive to the key-activities of today. Rural areas in the southern part of France for example, show a much more impressive economic performance than the older traditional manufacturing regions in the northern and eastern parts of the country. The same trend can be observed also in other countries (Keeble et al. 1983, Fothergill and Gudgin 1982).

This chapter will utilise the national line of reasoning followed by Aydalot and the spatial pattern of high-growth activities in The Netherlands will be examined. The main research question to be addressed is: does the contemporary pattern follow the traditional pattern of economic activities in The

Netherlands? or Do we see, as did Aydalot in France, a shift in the spatial distribution of the current 'key-activities'?

High Growth Activities

We are deliberately using the term 'high-growth' activity because from an economic point of view it is not as useful to classify an activity as high-tech. The main goal of an enterprise is to make a profit. Our interest in high-tech activities is, of course, related to expected market prospects for such activities. However, not all high-tech products are a commercial success, as the technologically very advanced Philips videorecorder (V 2000) illustrates. On the other hand, there are many activities that cannot be classified as high-tech, but do offer marvellous commercial opportunities for success. So The Netherlands Economic Institute (NEI 1984) made a list of 121 sub-sectors (four-digits level) with good commercial prospects. Of these 121 sub-sectors only 46 could be considered high-tech (38 per cent). Outside the manufacturing sector there is an especially large discrepancy between high-growth and high-tech. Only 6 out of 42 sectors with good commercial prospects could be considered high-tech where as for manufacturing 40 sub-sectors were high-tech out of 79 sub-sectors with good commercial prospects. This study is based on high-growth activities or on products that have reached a growth phase in their life cycle. After all it is the performance of firms, not the high-tech character of the activities in which they are involved that influences the spatial pattern of economic activities.

As in other West European countries we in Holland are trying to restructure our economy by stimulating activities for which we have or can develop a comparative advantage. Our most obvious comparative advantages are with those activities intrinsically connected with certain regions. But these advantages are also rather immobile. You either have beaches, gas, woods, ports etc. or you do not. It is hardly possible to change the situation in the short term. A second type of advantages can however, be influenced by policy: the number of universities, their economic functioning, the educational level of the labour force, connections with fore- and hinterland etc. In general these kinds of advantages are identified with Holland as a whole, more than with individual regions. This observation may be explained by the fact that The Netherlands is a small and densely populated country which does not manifest large interregional variations in these phenomena. The third kind of advantages are related to firms located within certain regions. True, there may at least have been an (historical) relation between the presence of a firm in a region and the locational characteristics of that region. Nevertheless, the presence of high-growth firms cannot always be seen as a feature intrinsic to that region. High growth firms can decide to change their location within Holland, and some do. So these firms are strong elements within but not necessarily of regions or countries.

The kind of advantages described above can be used to define sectors that are considered high-growth sectors for Holland as a whole and for certain regions in particular. The port of Rotterdam, the airport of Schiphol, and the beaches along the North Sea can be seen as region-based advantages. All or nearly all regions in The Netherlands have favourable locational conditions for skill- or R&D intensive activities. The activities of companies like Shell, Philips, Unilever, AKZO etc. in specific regions is the third element in this profile.

How can high-growth activities be identified? Our national government installed two advisory commissions of knowledgeable and distinguished citizens to carry out this task. These commissions followed an international line of reasoning in trying to define high-growth activities for The Netherlands. One of them (the Wagner Commission) considered the manufacturing sector, the other (the Oostenbrink Commission) studied business services activities. Both Commissions initially determined in which sectors Holland made a good showing in world trade. A good performance in a sector with high international growth-prospects was considered positively. This procedure can be also related to the portfolio approach (Wever and Grit 1984). It incorporates not only demand prospects but also the market position of the producer (in this case: The Netherlands). A strong market position in a growing sector that is not high-tech at all is more promising than a weak position (a minor market share) in a high-growth, high-tech sector. Additionally, activities with expected prosperous commercial futures, in the opinion of the experts, were identified by both commissions. This procedure was based on consensus and was strongly oriented to high-growth, high-tech activities (especially in manufacturing), that were expected to reach shortly the growth phase in their product life cycles. Based on both procedures a list of high-growth activities for The Netherlands has been established. This list was tested for the metal sector. In general high-growth branch-codes did indeed achieve a higher growth of exports, sales and investments than branch-codes that were less promising (Jansen and de Quilettes 1984).

A different procedure however, has recently been followed (for manufacturing only) by the Economic Institute of Small and Medium Sized Firms (EIM 1985). For the period 1978-1981 based on seven performance indicators (growth in inland sales, growth of exports, share of exports in total sales, growth in employment, growth in value added, growth of investments in fixed assets) all manufacturing sectors have been classified into six categories. These are: expansive, slightly stagnating, stagnating, saturating, slightly contracting and contracting. This classification, based on product life cycle reasoning, was tested for the period 1981-1983 and proved relatively stable.

These classifications will now be used to provide a picture

of the spatial pattern of high-growth activities in The Netherlands. However, the use of these classifications create two major
well-known problems. The first problem arises because our
research is not based on individual products. Indeed, the two
advisory commissions were perfectly right in classifying products
and not sectors. So for example, the production of robots was
considered by the commissions to be a promising activity.
However, our National Census does not distinguish, as a sector,
'robot production'. The offical classification is neither based
on individual products nor on products that will become important
in the future. In order to construct a spatial pattern we
therefore have to 'translate' and incorporate the list of
individual high-growth products into the aggregated Census-
classifications of high-growth sectors.

The second problem has to do with the branch-code given to
multi-product enterprises. If its main activity is not considered
a high-growth activity, an enterprise is not classified as such,
even though the portfolio-mix contains high-growth activities. To
give just one example: the branch-code often given to Philips-
Eindhoven is neither a high-growth, nor high-tech code in the NEI
1984 list.

The Data

The description of the spatial pattern of high-growth activities
in The Netherlands can be based either on the location of all
high-growth enterprises or on only new firms. Most firms
(existing and new ones) constantly make strategic decisions.
Sometimes the products they make reach the last phase of their
life cycle. To guarantee the continuity of the enterprise the
market position has to be strenghtened by reducing costs (using
cheaper raw materials or production processes) and/or searching
for new markets and/or introducing more efficient organisational
structures. In spite of such efforts the market position of a
firm may deteriorate. As a consequence it may be necessary to
adapt the product-mix, by introducing a new quality in one of the
existing products or even by introducing a totally new product.
In other words, many enterprises are constantly in search of new
high-growth activities that fit into their organisation.

In this respect there is a difference between existing and
potentially new firms. Most existing firms when changing their
product-mix do not change the location of their plants. This is
the inertia-factor, that we are familiar with in geography. An
inertia-factor may be noticed in new firms as well. We will deal
with this aspect later. It is nevertheless possible that it is
the location of new firms in high-growth activities that
illustrate spatial changes best. Moreover it is often assumed
that it is new firms or small firms in general, that are playing
the major role in the process of innovation and employment
creation (Rothwell and Zegveld 1982). Some evidence in support of

Table 2. Employment and number of firms in manufacturing per size-category as a percentage of total employment and number of firms in manufacturing

Phase in product cycle	Enterprises Small-sized		Medium-sized		Large-sized		Total	
	empl.	no	empl.	no	empl.	no	empl.	no
expansive	14.0	16.6	7.4	8.0	5.4	6.7	6.9	14.5
slightly stagnating	7.9	9.7	14.3	12.9	42.1	26.4	31.2	11.0
stagnating	48.5	41.9	42.6	44.2	36.3	40.6	39.3	42.3
saturated	10.9	12.3	12.8	13.1	5.3	10.2	7.9	12.4
slightly contracting	15.4	15.3	18.5	18.0	7.4	11.5	11.2	15.8
contracting	3.2	4.1	4.4	3.8	3.4	4.5	3.7	4.1
	100.0	100.0	100.0	100.0	100.0	100.0	100.0	100.0

Source: EIM 1985, 55 and 57

this position can be found in Table 2 in which the EIM-classification is related to firm size. Small firms show a over-representation in the category 'expansive firms'. Of course, expansive firms will influence strongly the spatial pattern of economic activities. However, in spite of these arguments, we may expect fairly small differences between the spatial patterns of the total existing and the new high-growth firms. Firstly, most entrepreneurs start their firm in the area where they live and in activities with which they are familiar. Secondly, spin-offs from existing firms are generally located near the incubator-organisation.

The data used about new firm formation in high-growth activities is based on the registration of firms by the Chambers of Commerce. By using this data-source nearly all new firms in the market segment have been included in this research project (Wever 1983). To reduce chance, to a certain extent, we used all new registrations for the two years, 1975 and 1980.

As spatial units we chose the districts used by the Chambers of Commerce. It would have been possible to construct a very detailed distribution pattern. But within the framework of this book this would not be appropriate. We will therefore, present only figures for a restricted number of combined districts, located all over the country (Fig. 1). The six regions so constructed vary widely in their characteristics. Two of them, the Amsterdam and Rotterdam areas, can be considered metropolitan areas. These are the 'poles' in the centre of the Dutch economic system. Two of the traditional Dutch problem areas have been included as well (the North and the South). The remaining regions are located in the southern part of what is called the Intermediate zone and in the periphery of the country.

Figure 1. The districts of the Chambers of Commerce and the classification centre-intermediate zone-periphery

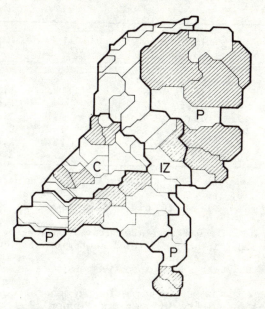

The Overall Picture

The overall picture of new firm formation in high-growth activities as defined by the two advisory commissions is given in Table 3. The figures are based on the slightly different 'translation' of high-growth activities in census sectors by two research institutes (TNO, see Alders and De Ruijter 1984 and NEI 1984). Both research projects include manufacturing as well as service sectors. The total number of new firms was established for each list. The mean number is related first to the total number of new firms and then to the number of inhabitants. Moreover a location quotient has been determined.

Looking first to the share of new high-growth firms in the total new firms, we see a low score for the problem areas (North and South). However, the discrepancy with the national average is remarkably small. In 1975 the South achieved a score of 71 per cent of the national figure, in 1980 the North came to 74 per cent. The difference is much larger however, in the number of new firms in high-growth sectors per 10,000 inhabitants where the South (1975) and North (1980) achieved only 48 and 50 per cent of the national total. This difference between both indicators is relevant. It is often assumed that locational conditions in peripheral areas are unfavourable for high-growth and/or high-

Table 3. New firm formation in high-growth activities in 1975 and 1980: all activities

Area	1975			1980		
	A	B	C	A	B	C
North	8.4	0.92	0.86	8.4	1.29	0.74
East	8.8	1.34	0.87	10.4	1.84	0.92
South	7.1	1.11	0.71	9.1	1.67	0.80
Intermediate	9.5	2.15	0.95	13.0	3.65	1.14
Rotterdam	12.9	2.82	1.28	13.1	3.52	1.15
Amsterdam	9.7	2.35	0.97	10.3	2.84	0.91
Netherlands	10.0	1.90		11.4	2.60	

A = number of new firms in high-growth sectors as a percentage of total number of new firms

B = number of new firms in high-growth sectors per 10.000 inhabitants

C = share in national number of new firms in high-growth sectors divided by share in national total number of all new firms

tech activities. For new firm formation in general we can refer to the incubation-theory (Hoover and Vernon 1959, Leone and Struyk 1976), for high-tech or high-growth activities to the product life cycle (Vernon 1966).

High-growth activities definitely are not considered 'footloose'. When being in the first phase of their life cycle they are supposed to flourish best in metropolitan areas, where they can best meet their need for information (Thompson 1969, Norton and Rees 1979). Peripheral areas on the other hand would offer a favourable environment for standardised products (Erickson 1976). If we use this approach we have to conclude from Table 1 that such a situation does not exist in The Netherlands. The number of high-growth firms created in the Dutch problem areas in the periphery is scarcely less, relatively, than in the country as a whole. Two explanations for this outcome can be given. It might be that high-growth activities are much more footloose than theory assumes. If we reject this idea we have to conclude that nowhere in The Netherlands do locational conditions prevent the setting up of such types of new firms. In the problem areas the main problem is the lack of new initiatives, and a lack of dynamism. It is not the 'wrong' firm that is set up there.

Another interesting conclusion concerns the two metropolitan areas. In line with theory the Rotterdam area (the southern wing of the Dutch Randstad) shows up as having a high firm formation rate. However, the scores of the Amsterdam region are much lower. In 1980 this area registered even less new firms in high-growth sectors per 10.000 inhabitants than the far less urbanised intermediate zone. Between 1975 and 1980 this intermediate zone has strenghtened its position. Together with the Rotterdam area this region has a location quotient above 1.00.

Figure 2. Spatial pattern of high-growth activities in The Netherlands (Jan. 1984): location quotients for all activities.

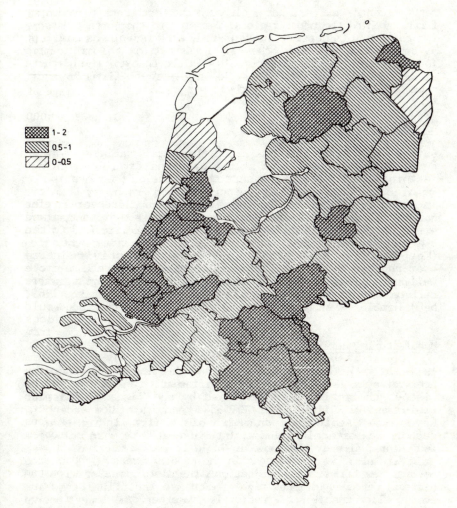

Source: Alders and De Ruijter (1984)

To what extent does this pattern of new firms coincide with the pattern for all firms in high-growth sectors? Fig. 2 presents location quotients (share in high-growth employment/share in total employment). The most striking outcome is again the small differences in the location quotients of the various regions. We see neither 'Silicon Valleys' nor 'centres d'excellence' near (technical) universities. There is not even a clear distinction High-growth activities in The Netherlands

between centre and periphery, although the North remains somewhat further behind the other regions. Compared to the North the score for the South is better. The Randstad, although not as dominant as centre-periphery approaches might suggest, is the main centre for existing high-growth firms. This is the case for the northern as well as the southern wings. New high-growth firms however, seem to be oriented to the southern wing.

A More Detailed Picture

Before drawing conclusions about the overal spatial pattern we should take a closer look at the patterns for the different sectors. After all we know that the choice of the location of manufacturing firms is influenced by factors other than the location of firms in the service sector. Moreover in The Netherlands as well as in other countries, to some extent, there exists a spatial specialisation. The economic centre (mostly the big cities) nowadays is strongly oriented to service activities (Keeble 1985, Bade 1984). Manufacturing activities are over-represented in the periphery, partly as the result of the well-known urban-rural shift in manufacturing. Althpugh a further differentiation within the manufacturing and service sector would be desirable, we will restrict ourselves to these two sectors.

Manufacturing activities

In Table 4 the spatial pattern of high-growth manufacturing activities is given. We used four different 'translations' of the list of the activities classified by the Wagner Commission (Alders and De Ruijter 1984, NEI 1984, Chambers of Commerce 1984, Wever 1984). Again the mean number of new firms in high-growth sectors was determined and with the use of this mean value the indicators, already mentioned in Table 1, were determined.

Here also, we notice that in the problem areas it is not the 'wrong' new firms that are 'set up'. The discrepancies with the national percentage of new high-growth manufacturing firms are even smaller than for all activities together. The lowest scores for the South and North problem areas are 82 per cent and 94 per cent of the national value in 1975 and 97 per cent and 84 per cent in 1980. Related to the number of inhabitants these figures for the North are 53 per cent and 57 per cent. This difference between manufacturing and all activities is not surprising. In The Netherlands manufacturing activities in general are fairly uniformly spread over the country. This holds for all manufacturing, for high-tech manufacturing activities and, as Table 4 shows, for new manufacturing firms in high-growth sectors. This is the case then even for firms that are assumed to need specific locational requirements.

A second point to notice is the weak position of the metropolitan area of Amsterdam. Contrary to theory, this region

Table 4. New firm formation in high-growth activities in 1975 and 1980: manufacturing

Area	1975			1980		
	A	B	C	A	B	C
North	1.99	.21	0.86	1.74	.27	0.82
East	1.92	.29	0.93	2.10	.37	0.90
South	1.73	.27	0.83	2.01	.37	1.03
Intermediate	2.89	.65	0.95	2.83	.79	1.20
Rotterdam	2.54	.55	1.36	2.92	.79	1.39
Amsterdam	1.51	.36	0.85	1.48	.41	0.91
Netherlands	2.11	.40		2.07	.47	

A = number of new firms in high-growth sectors in manufacturing as a percentage of total number of new firms
B = number of new firms in high-growth sectors in manufacturing per 10,000 inhabitants
C = share of national number of new firms in high-growth sectors in manufacturing divided by share in national number of all new firms in manufacturing

does not seem to be attractive to new high-growth manufacturing firms. For Dutch readers the difference between the Rotterdam and Amsterdam area will be no surprise. The southern wing of our Randstad (the same holds for the province of Noord-Brabant in which our intermediate zone is predominantly located) is much more clearly a manufacturing area. The northern wing has specialised in service activities. In other words, the high scores for the Rotterdam and intermediate area (this area again strengthened its position) are not unexpected results. However, the scores for the metropolitan area of Amsterdam were less predictable. These scores are low, even compared with the Dutch problem areas: this is another argument for being careful in using a centre-periphery approach within a small country like Holland.

Table 4 is based to a certain degree on high-tech

activities. High-tech was not a criterion used in the EIM classification based on economic performance. Figures for the EIM list are presented in Table 5. For practical reasons only values for growing (expansive and sligthly stagnating) and contracting (slightly and strongly contracting) classes have been given. The overall outcome fits in perfectly with our previous results. High-growing sectors are not absent in the Dutch problem areas (of course only in a relative sense). And again we see good figures for the Rotterdam area. The discrepancy with the Amsterdam area is even more striking if we consider the expansive and contracting sectors separately. In the Rotterdam area 3.3 per cent of all new firms (36.4 per cent of all new manufacturing firms) in 1980 were classified as expansive, against 0.5 (4.9 per cent) as strongly contracting. For Amsterdam the same percentages are 1.1 (14.8) and 1.4 (20.3). The only difference with Table 2 is the less strong position of the intermediate zone.

Table 5. Number of new firms in 1980 in expansive and slightly stagnating (A) sectors and in contracting sectors (B) as a percentage of all new firms and of all new manufacturing firms

Area	% of Total		% of Manufacturing	
	A	B	A	B
North	3.6	1.8	38.2	19.1
East	3.4	1.9	33.7	18.1
South	3.4	2.0	39.5	22.9
Intermediate	3.9	2.3	38.6	22.6
Rotterdam	4.3	0.9	47.0	9.1
Amsterdam	3.0	1.8	42.1	25.4
Netherlands	3.3	1.8	36.9	19.2

Service activities

We also considered the service sector. The same procedure was followed, based on two 'translations' of the list of high-growth activities of the Oostenbrink Commission (Alders and De Ruijter 1984, NEI 1984). The figures are given in Table 6.

The picture is different from that of the manufacturing sector. The most striking difference is the much stronger position of the Amsterdam metropolitan area in the service sector. However, there is now hardly any difference between the Rotterdam and Amsterdam areas. Nevertheless there is a relatively large descrepancy between the problem areas and the metropolitan areas in terms of their shares of service activities as compared with their shares of manufacturing activities. For the South the number of new high-growth service firms was only 55 per cent in 1975 and 67 per cent in 1980 (manufacturing 82 per cent and 97 per cent respectively) of the national value. This difference illustrates the stronger orientation of high-growth service sectors towards metropolitan environments. However metropolitan areas do not have a monopoly of new firms in high-growth service activities. This observation is illustrated by the example of the intermediate zone which between 1975 and 1980 strengthened its position. In 1980 this region had the highest score for each of the three indicators in Table 6.

Conclusion

The result of our exercise is more relevant than might seem at

Table 6. New firm formation in high-growth activities in 1975 and 1980: services

Area	1975			1980		
	A	B	C	A	B	C
North	5.27	0.56	0.84	5.53	0.85	0.75
East	5.31	0.81	0.85	6.68	1.18	0.92
South	3.45	0.54	0.55	4.96	0.91	0.67
Intermediate	4.73	1.07	0.79	7.85	2.20	1.08
Rotterdam	7.98	1.74	1.26	7.78	2.09	1.06
Amsterdam	7.28	1.76	1.15	7.44	2.12	1.02
Netherlands	6.23	1.18		7.36	1.68	

A = number of new firms in high-growth sectors in the service sector as a percentage of total number of new firms
B = number of new firms in high-growth sectors in the service sector per 10,000 inhabitants
C = share of national total of new firms in high-growth sectors in services divided by the share in national total of all new firms in services

first sight, especially when we relate it to theories like the product life cycle, incubation theory, trickling down processes, and even the centre-periphery approach. These theories - in their spatial form - assume a relationship between urbanisation and innovation (high-tech or high-growth). They implicitly assume that locational conditions in peripheral regions are unfavourable to high-growth or high-tech activities, in comparison with conditions in metropolitan areas. In general there is no need to dispute the relationship between urbanisation and innovation, as Clapp and Richardson (1985, 241) state: "The hypothesis that early phases of modern economic growth will be accompanied by the spatial concentration of economic activity in some, however defined, core region has been confirmed too many times to require any elaboration".

However, we should be aware of spatial scale. Talking about centre/periphery or innovation/urbanisation in a small country like Holland may be irrelevant in many economic activities. The whole of The Netherlands, perhaps with the exception of some regions in the North, appears to be one 'urban field' (Pred 1977). This concept 'urban field' can be seen as the spatial counterpart of the concept 'footloose'. Of course, even within an urban field conditions are not the same everywhere, but the (relative small) differences will hardly effect the financial performance of most individual firms. The spatial pattern of (new) high-growth manufacturing activities in The Netherlands gives evidence for this statement. These new firms are found everywhere in Holland, although they are not considered footloose. The locational requirements they need can be supplied nearly all over the country. Even for high-growth service firms the distinction between the strongly urbanised Randstad and the intermediate zone has already disappeared.

Our general conclusion therefore clearly contradicts popular thinking, even when we assume that for some activities within manufacturing or services the spatial pattern may be different. This conclusion becomes even more convincing if we include the 'closure' rate of new firms. This may sound peculiar, as Birch (1979) already found that regional differences in death rates among new firms are much smaller than in birth rates. Nevertheless we did find some differences (Table 7) even though they were also small.

Table 7. New firms and high-growth new firms founded in 1975 that had disappeared* by the end of 1982, as a percentage of all new firms/high growth new firms respectively founded in 1975

Area	High-growth firms		All firms
	Manufacturing	Services	
North	37.4	39.7	50.5
East	55.2	45.7	46.1
South	50.8	33.8	46.3
Intermediate	49.2	46.2	48.9
Rotterdam	79.7	62.8	60.7
Amsterdam	65.4	56.8	59.3
Netherlands	57.7	51.7	52.3

* Closures and out-migrations

There is a distinct trend in the 'closure' rates. The highest rates are found in the metropolitan areas, the lowest in the problem areas. As a result the difference between problem and metropolitan areas is reduced if we look at the number of surviving new firms. The already strong position of the intermediate zone becomes even stronger, because of relatively low 'closure' rates. More important however, is the fact that there is again no reason to assume that conditions in the Dutch problem areas would be unfavourable for high-growth activities in general.

Of course 'closure' rates do not say everything about the economic performance of new firms. The aspirations of new entrepreneurs in metropolitan areas may be higher than in problem areas, resulting in more closures as well as in more fast growing companies. Research (Bleumink et al. 1985) contradicts this idea. Firms registering a good performance (sales, employment growth) were over-represented in districts of the intermediate zone. Metropolitan districts did not clearly stand out from peripherally located districts. Here too, the Rotterdam area had better results than the Amsterdam area.

These arguments additionally support the thesis that The Netherlands can, to a large extent, be seen as one urban field. Moreover, this urban field is expanding spatially. The strong position of the southern wing of the intermediate zone (confirmed by other research, see Kok, Offerman and Pellenbarg 1984, Kleinknecht and Mouwen 1985) will in time result in a shift south- or southeastwards of the centre of gravity in the Dutch economic system. This undoubtly can be related to the southward shift of the economic centre of Western Europe, as illustrated by spatial developments in France (Aydalot 1984) and West Germany (Bade 1984).

How did the intermediate zone achieve this strong position for high-growth activities? At least two complementary explanations can be given. The first focusses on the reduced advantages of metropolitan environments for many economic activities (especially manufacturing, wholesaling and building), because of diseconomies of scale. Here we can refer to the production cost theory, arguing that manufacturing industry faces at present higher operating costs in metropolitan areas. Or we can refer to the constrained location theory, that is oriented exclusively to the impact of factory floorspace supply constraints (Keeble 1985). Often the physical moving of firms out of metropolitan areas is seen as a consequence of the reduced attractiveness of metropolitan areas. At the same time this migration illustrates the urban field idea.

However, a relationship between migration of firms out of metropolitan areas and the incubation theory and the product cycle theory can be seen. The Myrdalian-like spread effect assumes that the metropolitan area remains the main centre of new initiatives. The strong position of the intermediate zone in the Netherlands would therefore be the result of spatial spin-offs from the economic centre. No doubt spread or trickle down effects have positively contributed to the position of the intermediate zone. This is achieved directly, but it may also be done indirectly, for as Fothergill and Gudgin (1982) noticed, higher costs can also result in relative low growth rates for existing firms in metropolitan areas. Yet, there is an additional explanation that is perhaps more suitable for the present situation in Holland. Because of poor economic prospects the number of migrating firms has been reduced considerably. Moreover, our figures clearly show that the intermediate zone is characterised by high new firm formation rates. The vast majority of these new firms have been created by individual people living there. This points to a strong endogenous element.

In contemporary literature about new firm formation (Keeble and Wever 1986) a number of ideas can be found concerning this endogenous element. Some regions seem to have a 'climate' that is more positive for stimulating people to take (high-growth) economic initiatives. Here we can refer to Molle (1982), who makes a distinction between the accessibility economic actors have to ideas, innovations, finance and the receptivity of these actors to new ideas, innovations etc. There is a correspondence here with Allan Pred's behavioural matrix. The accessibility factor can also be related to the incubation theory and to the product life cycle, or in general to the possibilities that a particular environment offers. The receptivity factor is more strongly related to the characteristics of the economic actors. In explaining the dynamics in regional economics we should include both factors. The relevance of the characteristics of the economic actors can be illustrated by the new firm formation process.

The number of new firm initiatives is often related to the characteristics of the existing industrial structure. Areas with many small firms generally have high new firm formation rates. In small firms employees are assumed to receive more appropriate experience and training for entrepreneurship than in bigger firms. Because of traditional familiarity with entrepreneurship the climate in areas with many small firms can also be more positive for new initiatives. In the Dutch intermediate zone the share of small firms in the total number of firms is relatively high. In the problem areas small firms are under-represented.

There also may be regions in which the absolute number of potential entrepreneurs is low. A longterm selective out-migration of young people may structurally reduce the regional pool of potential entrepreneurs. Areas in Holland with high firm formation rates such as the intermediate zone are also characterised by large flows of people migrating in- and out. From England we know that the famous 'Cambridge phenomenon' is at least partly related to activities carried out by in-migrants (Keeble and Kelly 1986). The Dutch problem areas also lack dynamism in population movement.

The new firm formation process illustrates the importance of the 'social' factor in yet another way. New firms in high-growth or high-tech sectors in a country like Holland are not created where conditions are best, but where the entrepreneurs are living. New entrepreneurs nearly always start their firm in or near the place they live, mostly even in or near their homes (Cross 1981). For them there is no location choice problem. The main research question for geographers should be: why are people in some regions more inclined to start their own firm or why do certain regions have more potential entrepreneurs? Within an urban field there is no reason to relocate when the activities of new entrepreneurs prove to be a success.

Theoretically we can relate the previous remarks to the incubation theory. This theory deals with differences on a micro scale within an urban field. The original conception (Hoover and Vernon 1959) connected high new firm formation rates in certain neighbourhoods to characteristics (locational advantages) of those (micro) areas (low housing costs, face-to-face contacts, central location, etc.). Nowadays, in the explanation of such high new firm formation rates more emphasis is laid on the population or social factor (Jansen 1981, De Ruijter 1983). After all, no one will dispute that the traditional high interest in entrepreneurship in gypsy-camps is not solely the result of a favourable environment, but should also be attributed to the 'climate' there.

There seem to be arguments for including this 'climate' factor in the explanation of spatial patterns of economic activity, be it high-tech or traditional enterprises, more than has been done until now. This holds true especially for explanations on the spatial scale of an urban field, where for many activities - although not for all - the financial performance of the firm is not or is only marginally influenced by the (regional) location of the firm. Or, in terms used by Smith (1966), it seems the case that the spatial margins to profitability are large.

One last comment will be made about the Dutch problem areas. If we accept the idea of an urban field, how do we explain the relatively bad situation in these regions? It will be clear from the foregoing arguments that the backwardness of these regions nowadays cannot convincingly be explained by serious shortcomings in their physical environments. Our figures about new firm formation in high-growth sectors do not indicate clear disadvantages. That is not to say that there are no disadvantages. Of course, there is the problem of a small regional market, especially important for the service sector. Some areas lack specific intrinsic comparative advantages and so on. However, the poor position of these regions should also be attributed to a lack of dynamism within the regional population. This lack of dynamism is illustrated by low numbers of new and surviving firms per 10.000 inhabitants. Regional policy should stimulate this internal dynamism. One possibility could be a policy of stimulating the creation of new firms.

On the other hand, we have to be realistic. The traditional emphasis on an external solution for the regional problem is quite understandable. If it is necessary to realise a considerable reduction of the unemployment rate in problem areas

the relocation of a big firm is of more help than the setting up
of many new firms. Nevertheless there are arguments in favour of
an additional policy to stimulate internal regional dynamism.
Most of these arguments are connected with the negative effects
of external control. For further details we refer to Firn (1975)
and Wever (1980). Moreover, a longtime dependency on big
companies may reduce the number of initiatives in a region
(Storey 1982). This even may result in a life style or climate
that is opposed to innovation, to dynamism or to the creation of
new firms (Illeris 1986). But as the lack of dynamism in problem
areas did not come into being overnight, changing this situation
will be a lengthly matter.

Summary

Everywhere in The Netherlands new firms in high-growth sectors
are created. The metropolitan areas do not dominate this process
as strongly as theories sometimes suggest. When closure-rates are
included there is even a remarkably strong position for the less
urbanised intermediate zone. In the problem areas there are
created in a relative sense nearly as many new high-growth
manufacturing firms as in the other regions, but in these problem
regions fewer new high-growth firms are created in the service
sector. It is the low absolute number of new initiatives relative
to the total number of inhabitants that identifies a problem of
great concern to these problem areas.

From a study of the spatial pattern of new high-growth firms
it is concluded that for such firms nearly the whole of Holland
can be seen as an urban field. Even in the problem areas
locational conditions evidently do not restrict the creation and
survival of new high-growth firms, which in general, are not
considered to be 'footloose'.

Several conclusions can be drawn from these findings.
Apparently the Dutch problem areas do not have to contend with
serious shortcomings in their environment. The main problem is a
lack of internal dynamism. Dutch regional policy should
additionally be directed towards stimulating this latent internal
dynamism.

This notion of internal dynamism also provides another
reason for the strong position of the intermediate zone in the
Netherlands. In the long term, it may well contribute to a
southeastward shift in the centre of gravity of the Dutch
economic system.

References

Alders, B.C.M. and P.A. de Ruijter (1984) De ruimtelijke spreiding van kansrijke aktiviteiten in Nederland, TNO, Apel-
doorn/Delft

Aydalot, Ph. (1984) 'Questions for regional economy', Tijdschrift voor Economische en Sociale Geografie, 75, 4-14

Bade, F.J. (1984) Die funktionale Struktur der Wirtschaft und ihre raumliche Arbeidsteilung, IIm/IP 82-27, Wissenschaftszentrum, Berlin

Birch, D.L. (1979) The job generation process, MIT, Cambridge Mass.

Bleumink, P., G. de Groot, J. Bilderbeek and E. Wever (1985) Nieuwe ondernemingen en regio, Geografisch Instituut, Nijmegen

Clapp, J.M. and H.W. Richardson (1985) 'Technological change in informationprocessing industries and regional income differentials in developing countries', International Regional Science Review, 9, 241-256

Cross, M. (1981) New firm formation and regional development, Gower, Westmead, Farnborough

Duijn, J.J. van (1983) The long wave in economic life, Allen & Unwin, London

EIM (1985) Innovatie en het MKB. Deelrapport 1: plaatsbepaling van de Nederlandse industrie: expansie en groottestructuur, Zoetermeer

Erickson, R.A. (1976) 'The filtering down process: industrial location in a nonmetropolitan area', Professional Geographer, 28, 254-260

Firn, J.R. (1975) 'External control and regional development', Environment and Planning, A7, 393-414

Fothergill, S. and G. Gudgin (1982) Unequal growth: urban and regional employment change in the UK, Heinemann, London

Freeman, Chr. (ed.) (1984) Long waves in the world economy, Frances Pinter, London

Hoover, E.M. and R. Vernon (1959) Anatomy of a metropolis, the job changing distribution of people and jobs within the New York Metropolitan Region, Harvard University Press, Cambridge Mass.

Illeris, S. (1986) 'New firm creation in Denmark' In: D. Keeble and E. Wever (Eds.) New Firms and Regional Development in Europe, Croom Helm, London

Jansen, A.C.M. (1981) ''Inkubatie-Milieu': analyse van een geografisch begrip', Geografisch Tijdschrift, 4, 306-316

Jansen, E. and R. de Quillettes (1984) Zijn kansrijke bedrijven succesvoller dan andere? Een proefanalyse van de metaalsector, Min. Econ. Zaken/VVK, Woerden

Keeble, D. (1985) 'The changing spatial structure of economic activity and metropolitan decline in the United Kingdom' In: H.J. Ewers (Ed.) The future of metropolis, De Gruyter, Berlin

Keeble, D. and T. Kelly (1986) 'New firms and high-technology industry in the United Kingdom: the case of computer electronics' In: D. Keeble and E. Wever (Eds.) New Firms and Regional Development in Europe, Croom Helm, London

Keeble, D. and E. Wever (1986) New Firms and Regional
    Development in Europe, Croom Helm, London
Keeble, D., P.L. Owens and C. Thompson (1983) 'The urban-rural
    manufacturing shift in the European Community', Urban
    Studies, 20, 405-413
Kleinknecht, A. (1984) Innovation patterns in crisis and
    prosperity: Schumpeter's long cycle reconsidered, Free
    University, Amsterdam
Kleinknecht, A. and A. Mouwen (1985) 'Regionale innovatie: ver-
    schuiving naar de 'Halfwegzone'?' In: W.T.N. Molle (Ed.)
    Innovatie en regio, Staatsuitgeverij, Den Haag
Kok, J.A.A.M., G.J.D. Offerman and P.H. Pellenbarg (1984) 'The
    regional distribution of innovative firms in the
    Netherlands' In: M. de Smidt and E. Wever (Eds.) A profile
    of Dutch economic geography, Van Gorcum, Assen, 129-150
Leone, R.A. and R. Struyk (1976) 'The incubator hypothesis:
    evidence from five SMSAs', Urban Studies, 13, 325-331
Maddison, A. (1983) Ontwikkelingsfasen van het kapitalisme, Aula,
    Utrecht
Mensch, G. (1977) Das technologische Patt, Fischer, Frankfurt
Molle, W.T.M. (1982) Technological change and regional
    development in Europe, NEI, Rotterdam
NEI (1984) Technologische vernieuwing en regionale ontwikkeling
    in Nederland (TRANSFER), Rotterdam
Norton, R.D. and J. Rees (1979) 'The product cycle and the
    spatial decentralization of American manufacturing',
    Regional Studies, 13, 141-151
Pred, A. (1977) City-systems in advanced economies, Hutchinson,
    London
Rothwell, R. and W. Zegveld (1982) Innovation and the small and
    medium-sized firm, Frances Pinter, London
Ruijter, P.A. de (1983) 'De bruikbaarheid van het begrip 'incuba-
    tiemilieu'', Geografisch Tijdschrift, 17, 106-110
Schumpeter, J.A. (1939) Business cycles, McGraw-Hill, New York
Smith, D.M. (1966) 'A theoretical framework for geographical
    studies of industrial location', Economic Geography, 42, 95-
    113
Storey, D.J. (1982) Entrepreneurship and the new firm, Croom
    Helm, Beckenham
Thompson, W.R. (1969) Contemporary economic issues, Irwin,
    Homewood
Vernon, R. (1966) 'International investment and international
    trade in the product cycle', Quarterly Journal of Economics,
    80, 190-207
Wemelsfelder, J. (1985) 'Kan het ontstaan van nieuwe technolo-
    gieën worden beïnvloed?', Economisch-Statistische Berichten,
    70, 220-224
Wever, E. (1980) 'Multi-vestiging ondernemingen en regionale
    ontwikkeling', Economisch-Statische Berichten, 65, 785-790
Wever, E. (1983) 'Cohort analysis in economic geography', Tijd-
    schrift voor Economische en Sociale Geografie, 74, 217-223

Wever, E. (1984) Nieuwe ondernemingen in Nederland, Van Gorcum, Assen

Wever, E. and S. Grit (1984) 'Research methods and regional policy' In: M. de Smidt and E. Wever (Eds.) A profile of Dutch economic geography, Van Gorcum, Assen, 39-63

INDEX

Index

Index